Study Guide to Accompany

Essentials of Statistics
for the Behavioral Sciences
Third Edition

Frederick J. Gravetter
Larry B. Wallnau

BROOKS/COLE PUBLISHING COMPANY

I(T)P® An International Thomson Publishing Company

Pacific Grove • Albany • Belmont • Bonn • Boston • Cincinnati • Detroit • Johannesburg • London
Madrid • Melbourne • Mexico City • New York • Paris • Singapore • Tokyo • Toronto • Washington

Senior Assistant Editor: *Faith B. Stoddard*
Editorial Assistant: *Stephanie M. Andersen*
Production Editor: *Mary Vezilich*

For more information, contact:

BROOKS/COLE PUBLISHING COMPANY
511 Forest Lodge Road
Pacific Grove, CA 93950
USA

International Thomson Publishing Europe
Berkshire House 168-173
High Holborn
London WC1V 7AA
England

Thomas Nelson Australia
102 Dodds Street
South Melbourne, 3205
Victoria, Australia

Nelson Canada
1120 Birchmount Road
Scarborough, Ontario
Canada M1K 5G4

International Thomson Editores
Seneca 53
Col. Polanco
11560 México, D. F., México

International Thomson Publishing Japan
Hirakawacho Kyowa Building, 3F
2-2-1 Hirakawacho
Chiyoda-ku, Tokyo 102
Japan

International Thomson Publishing Asia
60 Albert Street
#15-01 Albert Complex
Singapore 189969

International Thomson Publishing GmbH
Königswinterer Strasse 418
53227 Bonn
Germany

Printed in Canada

5   4   3   2

ISBN 0-534-35806-3

CONTENTS

How to Use
This Study Guide

As the title indicates, the purpose of a Study Guide is to provide you with a guide for studying. In general, studying involves revisiting and reviewing material that you have already seen in class and in the textbook. Notice that we have used the word "review," as in "a second look." The textbook is written to provide an introduction to statistical concepts, and includes all the explanations and examples necessary to present new ideas that students are seeing for the first time. This Study Guide, on the other hand, is written to provide a second look at material that is already somewhat familiar. That is, the Study Guide is intended as a "review" that condenses, summarizes, and highlights the textbook. Thus, the Study Guide is a supplement to the textbook, not a replacement for the textbook, and we strongly suggest that you work in the Study Guide only after you have read and studied the corresponding section in the textbook.

Each chapter in the Study Guide corresponds to a chapter in the textbook, and each Study Guide chapter is divided into seven major sections. The following paragraphs provide an overview of these sections and some suggestions for how they should be used.

Chapter Summary. We begin with a big-picture overview of the chapter contents. The intent is to provide a relatively concise summary of the general content of the chapter and to identify most of the major points. This is the kind of thing that your mother wanted to know when she asked, "What did you learn in school today?" If you understand the big picture, then all the details, terminology, and formulas will be much easier to learn. Keep in mind, however, that the summary does not include all of the individual topics that are covered in the chapter. You will need to read the complete chapter and work some of the problems before you will pick up all the details.

<u>Learning Objectives</u>. Next we provide a list of goals that should be achieved upon completion of studying a text chapter. These objectives usually are task-oriented, pointing to specific things you should know how to do or how to explain. If you lack confidence in meeting any of these objectives, you should return to the appropriate section(s) of the text for further study.

<u>New Terms And Concepts</u>. This section consists of a list of important terms and concepts that appeared in the text. You should identify or define these terms, and explain how closely-related terms are interrelated. Return to the text to check your answers.

<u>New Formulas</u>. A list of formulas that were introduced in the text is provided. For each formula, you should test yourself in the following way:
1) Identify each symbol and term in the formula.
2) Make a list of the computational steps that are indicated by the formula.
3) Describe when each formula is used and explain why it is used or what it is computing.
4) Most importantly, you should realize that each formula is simply a concise, mathematical expression of a concept or procedure. If you can explain the concept or procedure <u>in words,</u> then it is usually very easy to translate the words into mathematical symbols and recreate the formula. Thus, we encourage you to understand concepts rather than memorize formulas.

<u>Step-By-Step</u>. This section presents a typical problem (or problems) from the chapter and provides a demonstration of the step-by-step procedures for solving the problem.

<u>Hints and Cautions</u>. In this section, we provide advice on the typical mistakes students make, and the difficulties they commonly have with the chapter material.

<u>Self-Test</u>. In this section we present a series of questions and problems that provide a general review of the chapter. As much as possible, the problems consist of simple sets of numbers so that you can solve them with minimal calculation. Work through the

Self-Test carefully, and remember that your mistakes will tell you what areas need additional study. Incidentally, the <u>Answers</u> to the Self-Test are given at the end of each Study Guide chapter. For problems with numerical answers, do not fret about slight differences between your answer and ours. A little rounding error is to be expected.

STUDY HINTS

It seems appropriate for a study guide to have a few suggestions to help you study. The following are some hints that have proved useful for our own students.

1. You will learn (and remember) much more if you study for short periods several times a week, rather than concentrating all your studying into one long session. For example, it is far more effective to study for one-half hour every night than to spend a single 3½ hour session once a week.

2. Do some work before class. Read the appropriate sections in the textbook before your instructor presents the material in class. Although you may not completely understand what you read, you will have a general idea of the topic which will make the lecture easier to follow. Also, you can identify items that are particularly confusing and then be sure that these items are clarified in class.

3. Pay attention and think during class. Although this may sound like obvious advice, many students spend their class time frantically taking notes rather than listening and understanding what is being said. For example, it usually is not necessary to copy down every sample problem that your instructor works in class. There are plenty of example problems in your textbook and in this study guide - you probably do not need one more in your notebook. Just put down your pencil and pay attention. Be sure that you understand what is being done, and try to anticipate the next step in the problem.

4. Test yourself regularly. Do not wait until the end of the chapter or the end of the week to check your knowledge. After each lecture, work some of the

end-of-chapter problems, do the learning checks, and be sure you can define key terms. If you are having trouble, get your questions answered immediately (re-read the text, go to your instructor, ask questions in class). Do not let yourself fall behind.

5. Don't kid yourself. Many students sit in class watching the instructor solve problems, and think to themselves, "This looks easy, I understand it." Do you really? Can you do the problem all by yourself?

Many students use the examples in the textbook as a guide for working assigned problems. They begin the problem, get stuck, then check the example to see what to do next. A minute later they are stuck again, so they take another peek at the example in the text. Eventually, the problem is finished and the students think that they understand how to solve problems. Although there is nothing wrong with using examples as models for solving problems, you should try working a problem with your book closed to determine whether you can complete it on your own.

Finally a few tips to help you prepare for exams.

1. Perhaps the best way to get ready for an exam is to make up your own exam. This is particularly effective if you have a friend in the class so you can both make up exams and then exchange them. Constructing your own exam forces you to identify the important points in the material, and it makes you think about how exam questions might be phrased. It is very satisfying to open a statistics exam and find some questions that you wrote yourself the day before.

2. Many students suffer from "exam anxiety" which causes them to freeze up and forget everything during exams. One way to help avoid the problem is for you to take charge of your own time during an exam.

a. Don't spend a lot of time working on one problem that you really don't understand (especially if it is a 1-point true/false question). Just move on to the rest of the exam - you can come back later, if you have time.

b. Remember that you do not have to finish the exam questions in the order they are presented. When you get your exam, go immediately to the problems you understand best.   This will build some confidence and make you better prepared for the remainder of the exam.

Probably the best way to reduce exam anxiety is to practice taking exams. Make up your own exam (select problems from the book and study guide) or have a friend make up an exam. Then try to duplicate the general conditions of an exam. If you are not allowed to use your book during exams, then put it away during practice. Give yourself a time limit. If you have an old alarm clock, set it in front of you so that you can hear the ticking and watch the time slip away. You might even try sitting in front of a mirror so that every time you look up there is someone watching you.

Finally, remember that very few miracles happen during exams. The work on your exam is usually a good reflection of your studying and understanding. Most students walk into an exam with a very good idea of how well they will do. Be honest with yourself. If you are well prepared, you will do well on the exam, and there is no reason to panic. If you are not prepared then exam anxiety probably is not your problem.

Chapter 1

Introduction to Statistics

CHAPTER SUMMARY

The general goals of Chapter 1 are:

1. To introduce the basic terminology that will be used in statistics.
2. To explain how statistical techniques fit into the general process of science.
3. To introduce some of the notation that will be used throughout the rest of the book.

Terminology

A <u>variable</u> is a characteristic or condition that can change or take on different values. Most research begins with a general question about the relationship between two variables for a specific group of individuals. The entire group of individuals is called the <u>population</u>. For example, a researcher may be interested in the relation between class size (variable 1) and academic performance (variable 2) for the population of third-grade children. Usually populations are so large that a researcher cannot examine the entire group, Therefore, a <u>sample</u> is selected to represent the population in a research study.

Variables can be classified as discrete or continuous. <u>Discrete</u> variables (such as class size) consist of indivisible categories, and <u>continuous variables</u> (such as time or weight) are infinitely dividable into whatever units a researcher may choose. For example, time can be measured to the nearest minute, second, half-second, etc.) To define the units for a continuous variable, a researcher must use <u>real limits</u> which are boundaries located exactly half-way between adjacent categories.

To establish relationships between variables, researchers must observe the variables and record their observations. This requires that the variables be <u>measured</u>. The process of measuring a variable requires a set of categories called a <u>scale of measurement</u> and a process that classifies each individual into one category. Four types of measurement scales are as follows:

a. A <u>nominal scale</u> is an unordered set of categories identified only by name. Nominal measurements only permit you to determine whether two individuals are the same or different.

b. An <u>ordinal scale</u> is an ordered set of categories. Ordinal measurements tell you the direction of difference between two individuals.

c. An <u>interval scale</u> is an ordered series of equal-sized categories. Interval measurements identify the direction and magnitude of a difference.

d. A <u>ratio scale</u> is an interval scale where a value of zero indicates none of the variable. Ratio measurements identify the direction and magnitude of differences and allow ratio comparisons of measurements.

Research studies can be classified as experiments, correlational studies, or quasi-experimental studies. In an <u>experiment,</u> one variable is manipulated and a second variable is observed to determine whether the manipulation causes changes. All other variables are controlled to prevent them from influencing the results. In an experiment, the manipulated variable is called the <u>independent variable</u> and the observed variable is the <u>dependent variable</u>. A <u>correlational</u> study simply observes the two variables as they exist naturally. A <u>quasi-experimental</u> study is similar to an experiment but is missing either the manipulation or the control necessary for a true experiment. In a quasi-experimental study the variable that differentiates the groups being compared is called a quasi-independent variable and is usually a pre-existing subject variable (such as male/female) or a time variable (such as before/after).

Statistics in Science

The measurements obtained in a research study are called the <u>data</u>. The goal of statistics is to help researchers organize and interpret the data. Statistical techniques are classified into two broad categories: descriptive and inferential. <u>Descriptive statistics</u> are methods for organizing and summarizing data. For example, tables or graphs are used to organize data, and descriptive values such as the average score are used to summarize data. A descriptive value for a population is called a <u>parameter</u> and a descriptive value for a sample is called a <u>statistic</u>. <u>Inferential statistics</u> are methods for using sample data to make general conclusions (inferences) about populations. Because a sample is typically only a part of the whole population, sample data provide only limited information about the population. As a result, sample statistics are generally imperfect representatives of the corresponding population parameters. The discrepancy between a sample statistic and its population parameter is called <u>sampling error</u>. Defining and measuring sampling error is a large part of inferential statistics.

Notation

The individual measurements or scores obtained for a subject will be identified by the letter X (or X and Y if there are multiple scores per subject). The number of scores in a data set will be identified by <u>N</u> for a population or <u>n</u> for a sample.

Summing a set of values is a common operation in statistics and has its own notation. The Greek letter sigma, Σ, will be used to stand for "the sum of." For example, ΣX identifies the sum of the scores. To use and interpret summation notation, you must follow the basic <u>order of operations</u> required for all mathematical calculation.

1. All calculations within parentheses are done first.
2. Squaring or raising to other exponents is done second.
3. Multiplying, and dividing are done third, and should be completed in order from left to right.
4. Summation with the Σ notation is done next
5. Any additional adding and subtracting is done last and should be completed in order from left to right.

For example, to compute $\Sigma X + 3$, you sum the X values, then add 3.
To compute $\Sigma(X + 3)^2$, you add 3 to each X (inside parentheses), then square the resulting values, then sum the squared numbers.

LEARNING OBJECTIVES

1. You should be familiar with the terminology and special notation of statistical methods.

2. You should understand the purpose of statistics: When, how, and why they are used.

3. You should understand summation notation and be able to use this notation to represent mathematical operations and to compute specified sums.

NEW TERMS AND CONCEPTS

The following terms were introduced in Chapter 1. You should be able to define or describe each term and, where appropriate, describe how each term is related to other terms in the list.

population	The entire group of individuals that a researcher wishes to study.

sample	A group selected from a population to participate in a research study.
statistic	A characteristic that describes a sample.
parameter	A characteristic that describes a population.
sampling error	The discrepancy between a statistic and a parameter.
descriptive statistics	Techniques that organize and summarize a set of data.
inferential statistics	Techniques that use sample data to draw general conclusions about populations.
variable	A characteristic that can change or take on different values.
constant	A characteristic that does not change.
raw score	An original, unaltered measurement.
dependent variable	In an experiment, the variable that is observed for changes.
independent variable	In an experiment, the variable that is manipulated by the researcher.
correlational method	A research method that simply observes the two variables being studied.

experimental method	A research method that manipulates one variable, observes a second variable for changes, and controls all other variables.
quasi-experimental method	A research method that compares groups of scores where the groups are defined by a non-manipulated variable.
control group	A group where the treatment is not administered.
experimental group	A group where the treatment is administered.
hypothetical construct	A characteristic or mechanism that is assumed to exist but cannot be observed or measured directly.
operational definition	A procedure for measuring and defining a construct.
nominal scale	A measurement scale where the categories are differentiated only by qualitative names.
ordinal scale	A measurement scale consisting of a series of ordered categories.
interval scale	An ordinal scale where all the categories are intervals with exactly the same width.
ratio scale	An interval scale where a value of zero corresponds to none.
discrete variable	A variable that exists in indivisible units.
continuous variable	A variable that can be divided into smaller units without limit.

Σ	Summation sign - the sum of
N	The number of scores in a population.
n	The number of scores in a sample.
X	A score.

STEP BY STEP

Summation Notation: In statistical calculations you constantly will be required to add a set of values to find a specific total. We will use algebraic expressions to represent the values being added (for example, X = score), and we will use the Greek letter sigma (Σ) to signify the process of summation. Occasionally, you simply will be adding a set of scores, ΣX. More often, you will be doing some initial computation and then adding the results. For example, we will routinely need to square each score and then find the sum of the squared values, ΣX^2. The following step-by-step process should help you understand summation notation and use it correctly to find appropriate totals.

Step 1: The first step in using summation correctly is to identify the "term" or "algebraic expression" that follows the summation sign. There are 3 general rules to help you identify the "term."
a. Everything contained within parentheses is part of the same term.
b. If several things are multiplied together, they are all part of the same term.
c. If something is squared, the squared sign is part of the term.

Step 2: Set up a computational table listing the original X values in the first column. Use the "term" you identified from Step 1 as a new column heading, and list all of the appropriate values for this term under the new heading. Suppose your task is to find $\Sigma(X + 3)$. The "term" in this expression is $(X + 3)$, so use this as a column heading and list all of the $(X + 3)$ values next to the original X values.

X	(X + 3)
4	7
8	11
2	5
6	9

Note: Occasionally you will need more than one column to get to the final term you want. To compute $\Sigma(X + 3)^2$, for example, begin with the X column, then add a column of $(X + 3)$ values, and then add a third column that squares each of the $(X + 3)$ values. Thus, each column represents a step in the computations.

X	(X + 3)	(X + 3)^2
4	7	49
8	11	121
2	5	25
6	9	81

Step 3: Simply add all the values in the column that is headed by the term identified in Step 1.

Using the same numbers that we used in Step 2, you find $\Sigma(X + 3)$ by simply adding the values in the $(X + 3)$ column.

$$\Sigma(X + 3) = 7 + 11 + 5 + 9 = 32$$

To find $\Sigma(X + 3)^2$ you add the values in the $(X + 3)^2$ column.

$$\Sigma(X + 3)^2 = 49 + 121 + 25 + 81 = 276.$$

HINTS AND CAUTIONS

1.   Many students confuse the independent variable and the dependent variable in an experiment. It may help you to differentiate these terms if you visualize an experiment as a big black box filled with people.

Independent Variable: The scientist (in a white lab-coat) stands outside the box and independently manipulates things (visualize knobs on the box that control things like temperature, background noise, etc.). The variable manipulated by the scientist is the independent variable.

Dependent Variable: Meanwhile, the purpose of the experiment is to see whether the people inside the box will respond to the scientist's manipulations. In other words, will the people's responses depend on what the scientist is doing. The dependent variable is the set of responses that the scientist observes and measures.

2.   There are three specific sums that are used repeatedly in statistics calculations. You should know the notation and computations for each of the following:
a. ΣX^2 First square each score, then add the squared values.
b. $(\Sigma X)^2$ First sum the scores, then square the total.
c. $\Sigma(X - C)^2$ First subtract the constant C from each score, then square each of the resulting values. Finally, add the squared numbers.

True/False Questions

1.   Statistical procedures that attempt to simplify and summarize data are classified as inferential statistics.

2.   Statistical procedures that use sample data as the basis for answering general questions about populations are called descriptive statistics.

3.   A characteristic (usually a single number) that describes a population is called a. parameter.

4.   A data set is described as consisting of $N = 15$ scores. Based on the notation being used, the data set is a sample.

5.   The average salary for a sample of $n = 20$ doctors would be an example of a parameter.

6.   One characteristic of an experiment is that the researcher manipulates one of the variables being examined.

7. Classifying a sample into two groups, males and females, is an example of measurement on a nominal scale.

8.   Classifying a sample of students according to high, medium, and low self-esteem would be an example of measurement on an interval scale.

9. To compute $\Sigma X - 1$, you first subtract one point from each score and then sum the resulting values.

10. When using summation notation, any operations inside parentheses are always performed first.

Multiple-Choice Questions

1. A research study typically begins with a question about a group of individuals. The entire group of individuals of interest is called the _____ .
  a. Parameter
  b. Statistic
  c. Population
  d. Sample

2. The relationship between a parameter and a statistic is the same as the relationship between a population and a(n) _____ .
  a. Variable
  b. Constant
  c. Score
  d. Sample

3. A researcher wants to examine the relationship between family size and political attitude for a group of 100 college students. Each student is asked to report the number of individuals in his/her immediate family and each student completes an attitude questionnaire that measures political opinions. What research method is being used in this study?
  a. Experimental method
  b. Correlational method
  c. Quasi-experimental method
  d. None of the above

4. A recent study reports that infant rats fed a special protein-enriched diet reached an adult weight 10% greater than litter-mates raised on a regular diet. For this study, what is the independent variable?

    a. The rats given the protein-enriched diet

    b. The rats given the regular diet

    c. The type of diet given to the rats

    d. The adult weight of the rats

5. A recent study reports that increased lighting during the winter months results in lower depression scores. For this study, what is the dependent variable?

    a. The amount of light

    b. The season of the year

    c. The level of depression

    d. The individuals who participated in the study

6. After measuring two individuals, a researcher can say that the score for one individual is larger than the score for the other, but it is impossible to say how much larger. What scale of measurement is being used in this situation?

    a. Nominal

    b. Ordinal

    c. Interval

    d. Ratio

7. Which of the following is an example of a discrete variable?

    a. Height

    b. Reaction time

    c. Number of brothers and/or sisters

    d. Age

8. What is the value of $\Sigma(X - 1)$ for the following scores?   Scores: 3, 4, 7.
   a. 11
   b. 12
   c. 13
   d. 14

9. What is the value of $(\Sigma X)^2$ for the following scores?   Scores: 1, 3, 5
   a. 36
   b. 81
   c. $(36)^2$
   d. 18

10. You are instructed to add 1 point to each score, then square the resulting values, and then find the sum of the squared numbers. In summation notation, this set of operations would be expressed as
   a. $(\Sigma X + 1)^2$
   b. $\Sigma(X + 1)^2$
   c. $\Sigma(X^2 + 1)$
   d. $\Sigma X^2 + 1$

Other Questions

1. Describe the relationships between a <u>sample</u>, a <u>population</u>, a <u>statistic</u> and a <u>parameter</u>.

2. What are the basic characteristics of an experiment that differentiate this method from other types of research?

3. Compute each value requested for the following set of scores.

X
1
3
5
2

ΣX = _____

ΣX^2 = _____

$(\Sigma X)^2$ = _____

N = _____

4. Compute each value requested for the following set of scores.

X
0
6
2
3

$\Sigma X + 1$ = _____

$\Sigma(X + 1)$ = _____

$\Sigma(X + 1)^2$ = _____

5. Use summation notation to express each of the following calculations.
   a. Add 3 points to each score, then find the sum of the resulting values.
   b. Find the sum of the scores, then add 10 points to the total.
   c. Subtract 1 point from each score, then square each of the resulting values. Next, find the sum of the squared numbers. Finally, add 5 points to this sum.

ANSWERS TO SELF TEST

True/False Answers

1. False. <u>Descriptive statistics</u> simplify and summarize data.

2. False. <u>Inferential statistics</u> use sample data to answer questions about populations.

3. True

4. False. The upper-case N indicates a population.

5. False.   The average for a sample is a _statistic_.

6. True

7. True

8. False. The categories (high, medium, low) form an ordinal scale.

9. False. Without parentheses, you first sum the scores, and then subtract 1.

10. True

Multiple-Choice Answers

1. c   2. d   3. b   4. c   5. c   6. b   7. c   8. a   9. b   10. b

Other Answers

1. A population is the entire group of individuals that you are interested in studying. A sample is a group selected from the population to participate in the research study. A parameter is a characteristic, usually a numerical value, of a population. A statistic is a characteristic of a sample.

2. The goal of an experiment is to establish a cause-and-effect relationship between two variables. To accomplish this goal, an experiment has two distinguishing features. First, the researcher manipulates one variable (the independent variable) to create the different treatment conditions that will be compared. The dependent variable is the score that is measured inside each treatment condition. The second feature of an experiment is that all other

variables are controlled so that they cannot influence the independent variable or the dependent variable.

3. $\Sigma X = 11, \ \Sigma X^2 = 39, \ (\Sigma X)^2 = 121, \ N = 4$

4. $\Sigma X + 1 \ = \ 11 + 1 \ = \ 12$
   $\Sigma(X + 1) = \ 1 + 7 + 3 + 4 \ = \ 15$
   $\Sigma(X + 1)^2 = \ 1 + 49 + 9 + 16 \ = \ 75$

5. a. $\Sigma(X + 3)$
   b. $\Sigma X + 10$
   c. $\Sigma(X - 1)^2 + 5$

Chapter 2

Frequency Distributions

CHAPTER SUMMARY

After collecting data, the first task for a researcher is to organize and simplify the data so that it is possible to get a general overview of the results. This is the goal of descriptive statistical techniques. One method for simplifying and organizing data is to construct a frequency distribution. Chapter 2 presents the general concept of a frequency distribution and explains several methods for constructing and displaying these distributions. In any given set of scores, we can determine how often a value is observed. "How often" is the frequency for that value. For example, a set of $N = 5$ scores consist of 3, 2, 2, 1, 2. For these data, the value $X = 2$ is observed three times. Therefore its frequency is $f = 3$. Frequency distributions are commonly presented in tables and graphs.

Frequency Distribution Tables: Frequency distribution tables contain at least two columns - one for X and another for f (frequency). In the X column values are listed from the highest to lowest, without skipping any. For the frequency column, tallies are determined for each value (how often each X value occurs in the data set. These tallies are the frequencies for each X value. The sum of the frequencies should equal N. A third column can be used for proportion (p). That is, we can determine what proportion of the entire data set consists a particular X value. For any X value,

$p = f/N$. The sum of the p column should equal 1.00. Finally, a fourth column displays the percentage of the distribution that any X value represents. The percentage is found by multiplying p by 100. The sum of this column, of course, should be 100%.

Sometimes a data set has a very wide range of values. In these situations a list of the X values in the first column is quite long - too long to provide enough simplification of the data. To remedy this situation, grouped frequency distribution tables are used. In the X column, class intervals are listed from highest to lowest interval, without skipping any interval. These intervals all have the same width, determined by the difference between the upper and lower real limits. The interval begins with a value that is a multiple of the interval width. The interval width (such as 2, 5, 10, 20, 50, 100) is selected so that the table will have approximately ten intervals.

Frequency Distributions Graphs: Frequency distribution graphs take the information from a frequency distribution table and essentially gives us a "picture" of the data set. The horizontal axis lists the values for X. The vertical axis is used for frequency and is labeled f. A histogram is a graph that uses bars. It is assumed that the X-axis is a continuous number line (reflecting measurement of a continuous variable). A bar is centered over each observed X value. The bar extends vertically to the corresponding frequency for that score. The width of the bar corresponds to the real limits of the X value. Thus, neighboring scores will have bars that touch.

A bar graph is like a histogram, except it is used with discrete variables. The bars for neighboring values or categories do not touch.

Polygons are graphs that may be used in place of histograms. Instead of using bars, a polygon places a dot centered above each X value. The vertical distance of the dot above X corresponds to the frequency for that score. The dots are then connected by straight lines and the lines are brought down to the X-axis at both ends of the graph.

Frequency distribution graphs are useful because they give us information about the shape of the distribution. A distribution is symmetrical if the left half of the graph is a mirror image of the right half. One example of a symmetrical distribution is the bell-shaped normal distribution. On the other hand, distributions are skewed when scores pile up on one side of the distribution, leaving a "tail" of a few extreme values on the other side. In a distribution with positive skew, the scores tend to pile up on the left side (low X values) of the distribution with the tail on the right. Negative skew is the opposite. In a negatively skewed distribution, most of the scores have high values (right side of the distribution) and the tail points to the left.

LEARNING OBJECTIVES

1. Know how to organize data into regular or grouped frequency distribution tables.

2. Be able to construct graphs, including bar graphs, histograms, and polygons.

3. Be able to describe the shape of a distribution portrayed in a frequency distribution graph.

NEW TERMS AND CONCEPTS

The following terms were introduced in this chapter. You should be able to define or describe each term and, where appropriate, describe how each term is related to other terms in the list.

frequency distribution   A tabulation of the number of individuals in each category on the scale of measurement.

grouped frequency distribution	A frequency distribution where scores are grouped into intervals rather than listed as individual values.
class interval	A group of scores in a grouped frequency distribution.
upper real limit	The boundary that separates an interval from the next higher interval.
lower real limit	The boundary that separates an interval from the next lower interval.
apparent limits	The score values that appear as the lowest score and the highest score in an interval.
histogram	A graph showing a bar above each score or interval so that the height of the bar corresponds to the frequency and width extends to the real limits.
bar graph	A graph showing a bar above each score or interval so that the height of the bar corresponds to the frequency. A space is left between adjacent bars.
polygon	A graph consisting of a line that connects a series of dots place above each score or interval. The height of each dot corresponds to the frequency.
symmetrical distribution	A distribution where the left-hand side is a mirror image of the right-hand side.

| positively skewed distribution | A distribution where the scores pile up on the left side and taper off to the right. |

| negatively skewed distribution | A distribution where the scores pile up on the right side and taper off to the left. |

| tail(s) of a distribution | A section on either side of a distribution where the frequency tapers down toward zero as the X values become more extreme. |

NEW FORMULAS

$$\text{proportion} = p = f/N$$

$$\text{percentage} = p(100) = (f/N)(100)$$

STEP BY STEP

Constructing a Frequency Distribution Table: The goal of a frequency distribution table is to take an entire set of scores and simplify and organize them into a form that allows a researcher to see at a glance the entire distribution. Suppose, for example, that an instructor gave a personality questionnaire measuring self-esteem to an entire class of psychology students. The questionnaire classifies each individual into one of five categories indicating different levels of self esteem: 1 = high self-esteem and 5 = low self-esteem.

The results for the class are as follows:

4, 4, 3, 3, 5, 4, 2, 1, 1, 3
4, 4, 5, 2, 3, 3, 4, 3, 3, 2
1, 4, 5, 3, 4, 4, 5, 2, 4, 1
3, 3, 2, 2, 4, 5, 1, 5, 3, 4

Step 1: In the first column of the table, list the scale of measurement starting with the highest score at the top and listing every possible X value down to the lowest score. The column heading should be "X" to indicate that this is the scale of X values.

X
5
4
3
2
1

Step 2: In a second column, headed by "f" for frequency, list the number of individuals who have each score. For example, six people have scores of $X = 5$, so you place a 6 in the f column beside the X value 5. Continue for each score (category) on the scale of measurement.

X	f
5	6
4	12
3	11
2	6
1	5

The result is a basic frequency distribution table. The table can be expanded by adding columns for proportion or percentage.

Retrieving Scores from a Frequency Distribution Table: Although a frequency distribution table provides a concise overview of an entire set of data, it is a condensed version of the actual data and for some students the table can obscure the details of the individual scores. For example, in the table we have just constructed, a set of 40 individual scores has been condensed into a table that shows only 10 numerical values (five X's and five frequencies). For some purposes, it is easier to transform a frequency distribution table back into a complete set of scores before you begin any statistical calculations with the data. We will use the table we have already constructed to demonstrate the process of recovering individual scores from a frequency distribution table.

Step 1: Find the number of individual scores (N): The frequency column (f) of the table shows the number of individuals located in each category on the scale of measurement. For this example, six individuals had scores of $X = 5$, twelve individuals had scores of $X = 4$, and so on. To determine the total number of individuals in the group, you simply add the frequencies.

$$N = \Sigma f = 6 + 12 + 11 + 6 + 5 = 40$$

Step 2: List the complete set of individual scores: Again, the table shows that six individual had scores of $X = 5$, twelve individuals had $X = 4$, and so on. These scores can be listed individually as follows:

5, 5, 5, 5, 5, 5	six fives
4, 4, 4, 4, 4, 4, 4, 4, 4, 4, 4, 4	twelve fours
3, 3, 3, 3, 3, 3, 3, 3, 3, 3, 3	eleven threes
2, 2, 2, 2, 2, 2	six twos
1, 1, 1, 1, 1	five ones

With the complete set of $N = 40$ scores listed in this way it is easy to perform computations based on individual scores. For example, to find ΣX you would add the 40 values ($\Sigma X = 128$).

1. When making the list of intervals for a grouped frequency distribution table, some people find it easier to begin the list with the lowest interval and work up to the highest. Fewer mistakes will be made.

2. When interpreting a frequency distribution table, be sure to use both columns, X and f, to get a complete list of the entire set of scores. Remember, the X column does not list all the scores, it simply shows the scale of measurement.

SELF-TEST

True/False Questions

1. By convention, a frequency distribution table lists the score categories (X values) from the highest score at the top to the lowest score at the bottom.

2. After the scores in a sample have been organized into a frequency distribution table, you can determine the number of scores (n) by simply adding the values in the f column of the table.

3. To find ΣX for the scores in a frequency distribution table, you simply add the values in the X column.

4. For a set of scores ranging from a high of X = 68 to a low of X = 61, you should use a grouped table instead of a regular frequency distribution table.

5. By convention, a grouped frequency distribution table is structured so that the number of intervals (the number of rows in the table) is a simple number such as 2, or 5, or 10.

6. A grouped frequency distribution table lists one interval as, 20-24. The width of this interval is 4 points.

7. In a frequency distribution graph, the scores (X values) are listed on the vertical axis.

8. If scores are measured on an interval scale, then you may use either a histogram or a polygon to display the frequency distribution.

9. A bar graph is constructed so that there are spaces between the bars.

10. In a negatively skewed distribution, the scores pile up on the right-hand side and taper off toward the left-hand side.

Multiple-Choice Questions

1. A sample of $n = 20$ scores ranges from a high of $X = 7$ to a low of $X = 3$. If these scores are placed in a frequency distribution table, how many rows will the table have?

     a. 4
     b. 5
     c. 7
     d. 20

Questions 2, 3, and 4 refer to the set of scores presented in the following table.

X	f
5	1
4	2
3	4
2	2
1	2

2. How many individual scores are in the entire set?
    a. $N = 5$
    b. $N = 6$
    c. $N = 11$
    d. Impossible to determine from the information given

3. How many individuals had a score of $X = 5$?
    a. 1
    b. 2
    c. 3
    d. 4

4. For this distribution, what is the value of ΣX?
    a. 5
    b. 11
    c. 15
    d. 31

5. A set of scores ranges from a high of $X = 67$ to a low of $X = 23$. If these scores are organized in a grouped frequency distribution table with an interval width of 5 points, then roughly how many intervals will by needed?

    a. 13-14

    b. 8-9

    c. 44

    d. 45

6. A researcher observes aggressive behavior for a sample of $n = 15$ boys and classifies each boy as high, medium, or low in terms of aggression. If the frequency distribution for these scores is presented in a graph, what kind of graph would be appropriate?

    a. A bar graph

    b. A histogram

    c. A polygon

    d. All of the above

Questions 7 and 8 refer to the following frequency distribution

X	f
24-25	2
22-23	4
20-21	6
18-19	3
16-17	0
14-15	1

7. For this distribution, how many individuals had scores lower than $X = 20$?

    a. 3

    b. 4

    c. 10

    d. Cannot be determined from the information given

8. For this distribution, the individual with the highest score has a value of
   a. X = 6
   b. X = 24
   c. X = 25
   d. Cannot be determined from the information given

9. A normal distribution is an example of what general shape?
   a. Symmetrical
   b. Positively skewed
   c. Negatively skewed.

10. Census data show that most people earn wages that are classfied as low to moderate wages, and only a few people earn wages that are high. Based on this information, the frequency distribution for income is
   a. negatively skewed
   b. positively skewed
   c. symmetrical

Other Questions

1. Simplifying and organizing data is the goal of descriptive statistics. One descriptive technique is to organize a set of scores in a frequency distribution. Define a frequency distribution and explain how it simplifies and organizes data.

2. Occasionally it is necessary to group scores into class intervals and construct a grouped frequency distribution.
   a. Explain when it is necessary to use a grouped table (as opposed to a regular table).
   b. Outline the guidelines for constructing a grouped frequency table.

3. A frequency distribution graph can be either a histogram, a bar graph, or a polygon. Define each of these graphs and identify the circumstances where each is used.

4. For a continuous variable each score actually corresponds to an interval on the scale of measurement.
   a. In general terms define the real limits of an interval.
   b. If a distribution has scores of 10, 9, 8, etc., what are the real limits for $X = 8$?
   c. If a distribution has scores of 5.5, 5.0, 4.5, 4.0, etc., what are the real limits for $X = 4.5$?
   d. In a grouped frequency distribution, each class interval has real limits and apparent limits. What are the real and apparent limits for the interval 10-14?

5. For the following set of scores:
   8, 7, 10, 12, 9, 11, 10, 9, 12, 11
   7, 9, 7, 10, 10, 8, 12, 7, 10, 7
   a. Construct a frequency distribution table including columns for frequency, proportion, and percentage.
   b. Draw a histogram showing these data.
   c. Draw a polygon showing the data.

ANSWERS TO SELF-TEST

True/False Answers

1. True

2. True

3. False. To find ΣX you must consider the f values as well as the X values.

4. False. It would require only 8 rows to list the full range of scores. A grouped table is not needed.

5. False. A grouped table should have about 10 intervals.

6. False. The interval width is 5 points.

7. False. The X values are listed on the horizontal axis.

8. True

9. True

10. True

Multiple-Choice Answers

1. b   2. c   3. a   4. d   5. b   6. a   7. b   8. d   9. a   10. b

Other Answers

1. A frequency distribution shows the number of individuals located in each category on the scale of measurement.

2. a. A grouped frequency distribution table is needed when the range of scores is large, causing a frequency distribution table to have to many entries in the X column.
    b. In a grouped frequency distribution table the guidelines are:
        1. You should strive for approximately 10 rows in the table.
        2. Interval widths of 2, 5, 10, 20, 50, and 100 should be used. (Select the interval width that satisfies guideline 1.) Interval width is determined by the difference between the real limits of the class interval.

3. The first (lowest) value of each interval should be a multiple of the interval width.
4. List all intervals without skipping any. The top interval should contain the highest observed X value and the bottom interval should contain the lowest observed X value.

3. In a histogram there is a bar above each score (or interval) showing the frequency. Adjacent bars are touching. A histogram is used with interval or ratio data. A bar graph is similar to a histogram except that there are spaces between the bars and the bar graph is used with nominal or ordinal data. In a polygon, the frequency is indicated by a dot above each score (or interval), and the dots are connected with straight lines. A polygon is used with interval or ratio data.

4.   a. The real limits for a score are the boundaries located halfway between the score and the next higher (or lower) score.
   b. The real limits for $X = 8$ would be 7.5 and 8.5.
   c. For $X = 4.5$, the real limits would be 4.25 and 4.75.
   d. For the class interval 10-14, the real limits are 9.5 and 14.5. The apparent limits are 10 and 14.

5. a.

X	f	p	%
12	3	.15	15%
11	2	.10	10%
10	5	.25	25%
9	3	.15	15%
8	2	.10	10%
7	5	.25	25%

b.

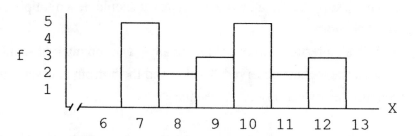

c.

Chapter 3

Central Tendency

CHAPTER SUMMARY

The purpose of Chapter 3 is to introduce the concept of central tendency and the three different statistical procedures that are used to define and measure central tendency. In general terms, central tendency is a statistical measure that determines a single value that accurately describes and represents an entire distribution of scores. The goal of central tendency is to identify the single value that is the best representative for the entire set of data.

By identifying the "average score," central tendency allows researchers to summarize or condense a large set of data into a single value. Thus, central tendency serves as a descriptive statistic because it allows researchers to describe or present a set of data in a very simplified, concise form. For example, the reading ability for an entire third-grade class can be summarized by the average reading score. In addition, it is possible to compare two (or more) sets of data by simply comparing the average score (central tendency) for one set versus the average score for another set. For example, a report may summarize research results by stating that the patients who received medication had an average cholesterol level 50 points lower than patients without medication.

It is essential that central tendency be determined by an objective and well-defined procedure so that others will understand exactly how the "average" value was obtained and can duplicate the process. Because no single procedure always produces a good, representative value, there are three commonly used techniques for measuring central tendency: the mean, the median, and the mode.

The Mean: The mean is the most commonly used measure of central tendency. Computation of the mean requires scores that are numerical values, usually measured on an interval or ratio scale. The mean is obtained by computing the sum, or total, for the entire set of scores, then dividing this sum by the number of scores.

For sample data the mean is: $\overline{X} = \dfrac{\Sigma X}{n}$

For population data the mean is: $\mu = \dfrac{\Sigma X}{N}$

Conceptually, the mean can also be defined as:
1. The mean is the amount that each individual receives when the total (ΣX) is divided equally among all N individuals.
2. The mean is the balance point of the distribution because the sum of the distances below the mean is exactly equal to the sum of the distances above the mean.

Because the calculation of the mean involves every score in the distribution, changing the value of any score will change the value of the mean. Also, modifying a distribution by discarding scores or by adding new scores will usually change the value of the mean. To determine how the mean will be affected for any specific situation you must consider: 1) how the number of scores is affected, and 2) how the sum of the scores is affected.

For example, adding a new score to a distribution will increase the number of scores by 1, and will increase ΣX by the value of the new score.

If a constant value is added to every score in a distribution, then the same constant value is added to the mean. Also, if every score is multiplied by a constant value, then the mean is also multiplied by the same constant value.

Although the mean is the most commonly used measure of central tendency, there are situations where the mean does not provide a good, representative value, and there are situations where you cannot compute a mean at all. When a distribution contains a few extreme scores (or is very skewed), the mean will be pulled toward the extremes (displaced toward the tail). In this case, the mean will not provide a "central" value. With data from a nominal scale it is impossible to compute a mean, and when data are measured on an ordinal scale (ranks), it is usually inappropriate to compute a mean. Thus, the mean does not always work as a measure of central tendency and it is necessary to have alternative procedures available.

The Median: The median is defined as the score or position in a distribution that divides the set of scores into two equal groups: exactly 50% of the scores are greater than the median, and exactly 50% are less than the median. Computation of the median requires scores that can be placed in rank order (smallest to largest) and are measured on an ordinal, interval, or ratio scale. Usually, the median can be found by a simple counting procedure:

1. With an even number of scores, list the values in order, and the median is the middle score in the list.
2. With an even number of scores, list the values in order, and the median is half-way between the middle two scores.

Often the simplest method for finding the median is to place the scores in a frequency distribution histogram with each score represented by a box in the graph. The goal is to draw a vertical line through the distribution so that exactly half the boxes are on each side of the line. The median is defined by the location of the line.

One advantage of the median is that it is relatively unaffected by extreme scores. Thus, the median tends to stay in the "center" of the distribution even when there are a few extreme scores or when the distribution is very skewed. In these situation, the median serves as a good alternative to the mean.

The Mode: The mode is defined as the most frequently occurring category or score in the distribution. In a frequency distribution graph, the mode is the category or score corresponding to the peak or high point of the distribution. The mode can be determined for data measured on any scale of measurement: nominal, ordinal, interval, or ratio.

It is possible for a distribution to have more than one mode. For example, a frequency distribution graph may have two peaks, with a mode at each peak. Such a distribution is called bimodal. (Note that a distribution can have only one mean and only one median.) In addition, the term "mode" is often used to describe a peak in a distribution that is not really the highest point. Thus, a distribution may have a major mode at the highest peak and a minor mode at a secondary peak in a different location.

The primary value of the mode is that it is the only measure of central tendency that can be used for data measured on a nominal scale. In addition, the mode often is used as a supplemental measure of central tendency that is reported along with the mean or the median.

Because the mean, the median, and the mode are all measuring central tendency, the three measures are often systematically related to each other. In a symmetrical distribution, for example, the mean and median will always be equal. If a symmetrical distribution has only one mode, the mode, mean, and median will all have the same value. In a skewed distribution, the mode will be located at the peak on one side and the mean will be displaced toward the tail on the other side. The median will be located between the mean and the mode.

1. You should be able to define central tendency and you should understand the general purpose of obtaining a measure of central tendency.

2. You should be able to define and compute each of the three basic measures of central tendency for a set of data.

3. You should know how the mean is affected when a set of scores is modified. For example, what happens to the mean when a new score is added to an existing set, or when a score is removed, or when the value of a score is changed. In addition, you should know what happens to the mean when a constant value is added to every score in a distribution, or when every score is multiplied by a constant value.

4. You should know when each of the three measures of central tendency is used and you should understand the advantages and disadvantages of each.

5. You should know how the three measures of central tendency are related to each other for symmetrical distributions and for skewed distributions.

6. You should be able to draw and understand graphs showing the relationship between an independent variable and a dependent variable, where a measure of central tendency (usually the mean) is used to present the "average" score for the dependent variable.

NEW TERMS AND CONCEPTS

The following terms were introduced in this chapter. You should be able to define or describe each term and, where appropriate, describe how each term is related to other terms in the list.

central tendency	A statistical measures that identifies a single score (usually a central value) to serve as a representative for the entire group.
mean	The value obtained when the sum of the scores is divided by the number of scores.
median	The score the divides a distribution exactly in half.
mode (major and minor)	The score with the greatest frequency overall (major), or the greatest frequency within the set of neighboring scores (minor).
weighted mean	The average of two means, where each mean is weighted by the number of scores it represents.

NEW FORMULAS

$$\mu = \frac{\Sigma X}{N}$$

$$\overline{X} = \frac{\Sigma X}{n}$$

The Weighted Mean: Occasionally a researcher will find it necessary to combine two (or more) sets of data, or to add new scores to an existing set of data. Rather than starting from scratch to compute the mean for the new set of data, it is possible to compute the weighted mean. To find the new mean you need two pieces of information:

  1) How many scores are in the new data set?
  2) What is the sum of all the scores?

Remember, the mean is the sum of the scores divided by the number. We will use the following problem to demonstrate the calculation of the weighted mean.

A researcher wants to combine the following three samples into a single group. Notice that sample 3 is actually a single score, $X = 4$. What is the mean for the combined group?

Sample 1	Sample 2	Sample 3
$n = 8$	$n = 5$	$n = 1$
$\overline{X} = 12$	$\overline{X} = 9$	$X = 4$

Step 1:   To find n and ΣX for the combined group, the first step is to find n and ΣX for each of the individual samples. For example, sample 1 consists of $n = 8$ scores with a mean of $\overline{X} = 12$. You can find ΣX for the scores by using the formula for the mean, and substituting the two values that you know,

$$\overline{X} = \frac{\Sigma X}{n}$$

In this case,

$$12 = \frac{\Sigma X}{8}$$

Multiplying both sides of the equation by 8 gives,

$$8(12) = \Sigma X$$
$$96 = \Sigma X$$

Often this process is easier to understand if you put dollar-signs on the numbers and remember that the mean is the amount that each individual receives if the total is divided equally. For this example, we have a group of 8 people (n) who have \$12 each ($\overline{X}$). If the group puts all their money together (ΣX), how much will they have? Again the answer is $\Sigma X = \$96$.

Step 2: Repeat the process in Step 1 for each individual data set. If the problem involves adding a single score to an existing data set, then you can treat the single score as a sample with $n = 1$ and $X = \overline{X} = \Sigma X$.

For this example, Sample 2 has a $n = 5$ and $\Sigma X = 45$, and Sample 3 has $n = 1$ and $\Sigma X = 4$.

Step 3: Once you have determined n and ΣX for each individual data set, then you simply add the individual n's to find the number of scores in the combined set. In the same way, you simply add the ΣX's to find the overall sum of the scores in the combined data set. For this example,

combined $n = 8 + 5 + 1 = 14$

combined $\Sigma X = 96 + 45 + 4 = 145$

Step 4: Finally you compute the mean for the combined group using the regular formula for $\overline{X}$.

$$\overline{X} = \Sigma X/n = 145/14 = 10.36$$

HINTS AND CAUTIONS

1. One of the most common errors in computing central tendency occurs when students are attempting to find the mean for data in a frequency distribution table. You must remember that the frequency distribution table condenses a large set of scores into a concise, organized distribution; the table does not list each of the individual scores. One way of avoiding confusion is to transform the frequency distribution table back into the original list of scores. For example, the following frequency distribution table presents a distribution for which three individuals had scores of $X = 5$; one individual had $X = 4$; four individuals had $X = 3$; no one had $X = 2$; and two individuals had $X = 1$. When each of these scores is listed individually, it is much easier to see that $N = 10$ and $\Sigma X = 33$ for these data.

Frequency Distribution   Original X Values

X	f
5	3
4	1
3	4
2	0
1	2

Original X Values

5
5
5
4
3
3
3
3
1
1

2. Many students incorrectly assume that the median corresponds to the midpoint of the range of scores. For example, it is tempting to say that the median for a 100-point test would be X = 50. Be careful! The correct interpretation is that the median divides the set of scores (or individuals) into two equal groups. On a 100-point test, for example, the median could be X = 95 if the test was very easy and 50% of the class scored above 95. You must know where the individual scores are located before you can find the median.

Often is easy to locate the median if you sketch a histogram of the frequency distribution. If each score is represented by a "block" in the graph, you can find the median by positioning a vertical line so that it divides the blocks into two equal piles.

SELF-TEST

True/False Questions

1. A sample of n = 6 scores has a mean of $\overline{X}$ = 12. For this sample ΣX = 72.

2. It is possible to have a distribution of scores where no individual has a score exactly equal to the mean.

3. If 5 points are added to every score in a sample, then the sample mean will not be changed.

4. A sample has a mean of $\overline{X}$ = 40. If a score of X = 55 is removed from the sample, then the sample mean would be increased.

5. On a 100-point test, you have a score of X = 73. Based on this information, you know that you definitely scored above the median.

6. A sample of n = 5 scores has a mean of 50. Another sample has n = 10 scores and a mean of 60. If the two samples are combined, the combined sample mean will be greater than 55.

7. A sample of n = 4 scores has a mean of $\overline{X}$ = 10. If a new score with a value of X = 5 is added to the sample, then the new sample will have a mean of 15.

8. It is possible for a distribution to have more than one mode.

9. For any symmetrical distribution, the mean, the median, and the mode will all have the same value.

10. A distribution of scores has a mean of 84 and a median of 80. Based on this information it appears that the distribution is positively skewed.

Multiple-Choice Questions

1. For the population of scores shown in the frequency distribution table, the mean is
   a. 15/5 = 3
   b. 15/12 = 1.25
   c. 34/5 = 6.80
   d. 34/12 = 2.83

X	f
5	2
4	1
3	4
2	3
1	2

2. Changing the value of a score in a distribution will <u>always</u> change the value of the

_____.

     a. Mean

     b. Median

     c. Mode

     d. All of the above

3. It is impossible to have a distribution where no one has a score exactly equal to the

_____.

     a. Mean

     b. Median

     c. Mode

     d. All of the above

4. A researcher measures personality for a sample of $n = 50$ people and classifies each person as either Type A, Type B, or Type C personality. Which measure of central tendency would be appropriate to summarize these measurements?

     a. Mean

     b. Median

     c. Mode

     d. Any of the three measures could be used

5. A distribution of scores has a mean of $\mu = 50$. One new score is added to the distribution and the new mean is found to be $\mu = 52$. From this result you can conclude that the new score was

     a. Greater than 50

     b. Less than 50

     c. Equal to 52

     d. Cannot answer from the information given

6. One sample has n = 10 scores and $\overline{X}$ = 30. A second sample has n = 20 scores and $\overline{X}$ = 40. If the two samples are combined, then the combined sample will have a mean
   a. Half-way between 30 and 40.
   b. Closer to 30 than it is to 40
   c. Closer to 40 than it is to 30
   d. None of the above

7. What is the value of the median for the following set of scores?
   a. 5
   b. 5.5   Scores: 1, 3, 3, 5, 6, 7, 8, 23
   c. 6
   d. 54/8 = 7

8. A population with a mean of μ = 8 has ΣX = 40. How many scores are in the population?
   a. N = 5
   b. N = 320
   c. N = 8/40 = 1/5
   d. Cannot be determined from the information given

9. Which of the following is true for a symmetrical distribution?
   a. The mean, median, and mode are all equal
   b. mean = median
   c. mean = mode
   d. median = mode

10. In a positively skewed distribution, which measure of central tendency will have the largest value?
    a. Mean
    b. Median
    c. Mode
    d. Impossible to determine from the information given

1. Explain the general purpose for obtaining a measure of central tendency.

2. A sample of $n = 6$ scores has a mean of $\overline{X} = 10$.
    a.   If a new score, $X = 3$, is added to the sample, what value will be obtained for the new sample mean?
    b.   If one of the scores, $X = 20$, is removed from the original sample, what value would be obtained for the new sample mean?
    c.   If one of the scores in the original sample is changed from $X = 6$ to $X = 24$, what value would be obtained for the new sample mean?

3. One sample has $n = 5$ and $\overline{X} = 20$. A second sample has $n = 15$ scores with $\overline{X} = 10$. If the two samples are combined, what is the value of the mean for the combined sample?

4. Compute the mean, median, and mode for the following set of scores.
    Scores: 5, 7, 5, 4, 3, 12, 9, 6, 6, 5, 7, 5, 6, 4

5. Compute the mean, median, and mode for the set of scores shown in the following frequency distribution table.

X	f
7	1
6	1
5	1
4	1
3	4
2	3
1	1

6. A researcher obtains the following results from an experiment comparing three treatment conditions:

Treatment #1: $\overline{X} = 12.7$
Treatment #2: $\overline{X} = 20.5$
Treatment #3: $\overline{X} = 8.4$

a. Assuming that the independent variable (the differences between treatments) is measured on a nominal scale, sketch a graph showing the experimental results.

b. Assuming that the independent variable is measured on an interval scale, sketch a graph showing the results.

ANSWERS TO SELF-TEST

True/False Answers

1. True

2. True

3. False. Adding 5 points to each score would add 5 points to the mean.

4. False. Removing a large score will cause the mean to decrease.

5. False (not necessarily true). You need to know where the individual scores are located to determine the median.

6. True

7. False. The new score is relatively small compared to the others, and would cause the mean to decrease.

8. True

9. False. The mean and median are equal but the mode may be different (for example, when there are two modes).

10. True

Multiple-Choice Answers

1. d   2. a   3. c   4. c   5. a   6. c   7. b   8. a   9. b   10. a

Other Answers

1. The purpose of central tendency is to find a single value that best represents an entire distribution of scores.

2. a. With n = 6 and $\overline{X}$ = 10, the original sample has a total of ΣX = 60. Adding X = 3 produces n = 7 and ΣX = 63. The new mean is 63/7 = 9.
   b. With n = 6 and $\overline{X}$ = 10, the original sample has a total of ΣX = 60. Taking away X = 20 leaves a sample with n = 5 scores and ΣX = 40. The new mean is 40/5 = 8.
   c. Changing X = 6 to X = 24 adds 18 points to the total but does not change the number of scores. The new sample has n = 6 with ΣX = 78. The new mean is 78/6 = 13.

3. The first sample has n = 5, $\overline{X}$ = 20, and ΣX = 100. The second sample has n = 15, $\overline{X}$ = 10, and ΣX = 150. When the samples are combined, the total number of score is n = 20 and the sum of the scores is ΣX = 250. The mean for the combined sample is 250/20 = 12.5.

4. The mean is 84/14 = 6.00. The median is X = 5.5, and the mode is X = 5.

5. The mean is 41/12 = 3.42. The median is X = 3.00, and the Mode is X = 3.

6. a.

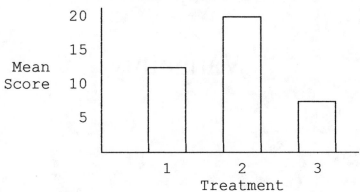

b. Use a histogram or a line graph (shown below).

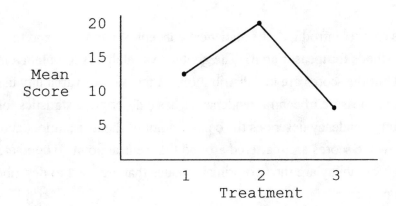

Chapter 4

Variability

CHAPTER SUMMARY

This chapter introduces the statistical concept of variability and the different statistical methods for measuring it. The goal for variability is to obtain a measure of how spread out the scores are in a distribution. A measure of variability usually accompanies a measure of central tendency as basic descriptive statistics for a set of scores. Central tendency describes the central point of the distribution, and variability describes how the scores are scattered around that central point. Together, central tendency and variability are the two primary values that are used to describe a distribution of scores.

Variability serves both as a descriptive measure and as an important component of most inferential statistics. As a descriptive statistic, variability measures the degree to which the scores are spread out or clustered together in a distribution. In the context of inferential statistics, variability provides a measure of how accurately any individual score or sample represents the entire population. When the population variability is small, all of the scores are clustered close together and any individual score or sample will necessarily provide a good representation of the entire set. On the other hand, when variability is large and scores are widely spread, it is easy for one or two extreme scores to give a distorted picture of the general population.

Variability can be measured with the range, the interquartile range, or the standard deviation/variance. In each case, variability is determined by measuring <u>distance</u>. The range is the total distance covered by the distribution, from the highest score to the lowest score (using the upper and lower real limits of the range). The interquartile range is the distance covered by the middle 50% of the distribution (the difference between Q1 and Q3). Standard deviation measures the standard distance between a score and the mean. The calculation of standard deviation can be summarized as a four-step process:

    a. Compute the deviation (distance from the mean) for each score.

    b. Square each deviation.

    c. Compute the mean of the squared deviations. For a population, this involves summing the squared deviations (sum of squares, SS) and then dividing by N. The answer value is called the <u>variance</u> or <u>mean square</u> and measures the average squared distance from the mean.

    d. For samples, variance is computed by dividing the sum of the squared deviations (SS) by n - 1, rather than N. The value, n - 1, is know as degrees of freedom (df) and is used so that the sample variance will provide an unbiased estimate of the population variance.

    e. Finally, take the square root of the variance to obtain the standard deviation.

LEARNING OBJECTIVES

1. Understand the measures of variability and be able to tell the difference between sets of scores with low versus high variability.

2. Know how to calculate SS using either the computational or definitional formula.

3. Be able to calculate the population and sample variance and standard deviation, and understand the correction used in the formulas for the sample statistics.

4. Be familiar with the characteristics of measures of variability, especially those for standard deviation.

NEW TERMS AND CONCEPTS

The following terms were introduced in this chapter. You should be able to define or describe each term and, where appropriate, describe how each term is related to other terms in the list.

variability	A measure of the degree to which the scores in a distribution are clustered together or spread apart.
range	The distance from the upper real limit of the highest score to the lower real limit of the lowest score; the total distance from the absolute highest point to the lowest point in the distribution.
first quartile	The score that separates the lowest 25% of a distribution from the highest 75%.
third quartile	The score that separates the lowest 75% of a distribution from the highest 25%.
semi-interquartile range	The distance from the first quartile to the third quartile.

deviation score	The distance (and direction) from the mean to a specific score. Deviation $= X - \mu$.
population variance	The average squared distance from the mean; the mean of the squared deviations.
population standard deviation	The square root of the population variance; a measure of the standard distance from the mean.
sample variance	The sum of the squared deviations divided by $df = n - 1$. An unbiased estimate of the population variance.
sample standard deviation	The square root of the sample variance.
unbiased estimate	An estimate that on average is right on the mark (not consistently too high or too low).
degrees of freedom	The number of scores in a sample that are free to vary with no restriction. $df = n - 1$.

NEW FORMULAS

For a population:

$$SS = \Sigma(X - \mu)^2 \quad \text{or} \quad SS = \Sigma X^2 - \frac{(\Sigma X)^2}{N}$$

$$\sigma^2 = \frac{SS}{N} \qquad \sigma = \sqrt{\frac{SS}{N}}$$

For a sample:

$$SS = \Sigma(X - \overline{X})^2 \quad \text{or} \quad SS = \Sigma X^2 - \frac{(\Sigma X)^2}{n}$$

$$s^2 = \frac{SS}{n-1} \qquad s = \sqrt{\frac{SS}{n-1}} \qquad df = n - 1$$

STEP BY STEP

<u>SS, Variance, and Standard Deviation</u>: The following set of data will be used to demonstrate the calculation of SS, variance, and standard deviation.

Scores: 5, 3, 2, 4, 1

Step 1: Before you begin any calculation, simply look at the set of scores and make a preliminary estimate of the mean and standard deviation. For this set of data, it should be obvious that the mean is around 3, and most of the scores are within one or two points of the mean. Therefore, the standard deviation (standard distance from the mean) should be about 1 or 2.

Step 2: Determine which formula you will use to compute SS. If you have a relatively small set of data and the mean is a whole number, then use the definitional formula. Otherwise, the computational formula is a better choice.

For this example there are only 5 scores and the mean is equal to 3. The definitional formula would be fine for these scores.

Step 3: Calculate SS. Note that it does not matter whether the set of scores is a sample or a population when you are computing SS. For this example, we will use formulas with population notation, but using sample notation would not change the result.

Definitional Formula: List each score in a column. In a second column put the deviation score for each X value. (Check that the deviations add to zero). In a third column list the squared deviation scores. Then simply add the values in the third column.

X	$(X - \mu)$	$(X - \mu)^2$
5	2	4
3	0	0
2	-1	1
4	1	1
1	-2	4
	0	10 = SS

Computational Formula: List each score in a column. In second column list the squared value for each X. Then find the sum for each column. These are the two sums that are needed for the computational formula.

$$\begin{array}{ll} X & X^2 \\ 5 & 25 \\ 3 & 9 \quad\quad \Sigma X = 15 \\ 2 & 4 \\ 4 & 16 \quad\quad \Sigma X^2 = 55 \\ \underline{1} & \underline{1} \\ 15 & 55 \end{array}$$

Then use the two sums in the computational formula to calculate SS.

$$SS = \Sigma X^2 - \frac{(\Sigma X)^2}{N}$$

$$= 55 - (15)^2/5 = 55 - 45$$

$$= 10$$

Step 4: Now you must determine whether the set of scores is a sample or a population. With a population you use N in the formulas for variance and standard deviation. With a sample, use n - 1.

For a Population   For a Sample

$$\sigma^2 = SS/N \quad\quad\quad\quad s^2 = SS/(n-1)$$
$$= 10/5 = 2 \quad\quad\quad\quad = 10/4 = 2.5$$

$$\sigma = \sqrt{2} = 1.41 \quad\quad\quad s = \sqrt{2.5} = 1.58$$

HINTS AND CAUTIONS

1. Mistakes are commonly made in the computational formula for SS. Often, ΣX^2 is confused with $(\Sigma X)^2$. The ΣX^2 indicates that the X values are first squared, then summed. On the other hand, $(\Sigma X)^2$ requires that the X values are first added, then the sum is squared.

2. Remember, it is impossible to get a negative value for SS because, by definition, SS is the sum of the squared deviation scores. The squaring operation eliminates all of the negative signs.

3. The computational formula is usually easier to use than the definitional formula because the mean usually is not a whole number.

4. When computing a variance or a standard deviation, be sure to check whether you are computing the measure for a population or a sample. Remember, the sample variance and standard deviation use n - 1 in the denominator so that these values will provide unbiased estimates of the corresponding population parameters.

5. Note that you do not use n - 1 in the formula for sample SS. The value n - 1 is used to compute sample variance and standard deviation after you have calculated SS.

True/False Questions

1. If a deviation score has a negative (-) sign, then the X value is below the mean.

2. A deviation score with a high numerical value (either positive or negative) indicates a score that is relatively close to the mean.

3. It is impossible to obtain a negative value for SS.

4. If a population has a variance of 4, then the population standard deviation is 16.

5. If 10 points are added to every score in a distribution, then the value of the standard deviation will not be changed.

6. Without some correction to the formula, sample variance is biased because it tends to overestimates the population variance.

7. A sample of n = 10 scores has SS = 90. For this sample, the variance is 9.

8. A population of N = 10 scores has SS = 90. For this population, the variance is 9.

9. The calculation of SS does not depend on whether a set of scores is viewed as a sample or a population.

10. Degrees of freedom (df) are used in the denominator of the formula for sample variance to obtain an unbiased estimate of population variance.

1. For any set of data, the sum of the deviation scores will always be
    a. Greater than zero
    b. Equal to zero
    c. Less than zero
    d. Impossible to determine without more information

2. What is the value of SS for the following sample?   Scores: 1, 3, 5
    a. SS = 8
    b. SS = 8/2
    c. SS = 8/3
    d. SS = $(8)^2$

3. Scores from a statistics exam are reported as deviation scores. Which of the following deviation scores indicates a higher position in the class distribution?
    a. +8
    b. 0
    c. -8
    d. Cannot determine without more informatiion

4. A population of N = 10 scores has μ = 50 and SS = 200. For this population, what is the value of $\Sigma(X - \mu)$?
    a. 0
    b. $\sqrt{200}$
    c, 450
    d. Cannot be determined from the information given

5. A population of N = 10 scores has μ = 50 and SS = 200. For this population, what is the value of $\Sigma(X - \mu)^2$?

    a. 0

    b. 200

    c, $(450)^2$

    d. Cannot be determined from the information given

6. Without doing any serious calculations, which of the following samples has the largest variance?

    a. 1, 3, 4, 5, 6

    b. 1, 5, 8, 12, 22

    c. 30, 32, 34, 35, 36

7. A population of scores has μ = 50 and σ = 10. If 5 points are added to every score in the population, then the new mean and standard deviation would be

    a. μ = 50 and σ = 10

    b. μ = 55 and σ = 10

    c. μ = 50 and σ = 15

    d. μ = 55 and σ = 15

8. A population of scores has μ = 50 and σ = 10. If every score in the population is multiplied by 2, then the new mean and standard deviation would be

    a. μ = 50 and σ = 10

    b. μ = 100 and σ = 10

    c. μ = 50 and σ = 20

    d. μ = 100 and σ = 20

9. If all n = 10 individuals in a sample have exactly the score, then the sample standard deviation will have a value of
   a. 0
   b. 10
   c. The standard deviation will equal the score
   d. Cannot determine from the information given

10. On an exam with a mean of $\mu = 70$, you have a score of X = 75. Which of the following values for the standard deviation would give you the highest position within the class?
    a. $\sigma = 1$
    b. $\sigma = 5$
    c. $\sigma = 10$
    d. Cannot determine from the information given

Other Questions

1. Calculate the variance and standard deviation for the following sample of scores: 3, 3, 6, 8, 2, 6, 7, 5.

2. Compute the variance and standard deviation for the following population of scores: 8, 9, 5, 2, 6.

3. Calculate SS, variance, and standard deviation for the following sample of scores, 15, 16, 9, 1, 9.

4. Describe what happens to the deviation scores and the standard deviation when a constant is added to every score in the distribution.

ANSWERS TO SELF-TEST

True/False Answers

1. True

2. False. A large deviation indicates a location far from the mean.

3. True

4. False. The standard deviation is the square root of the variance.

5. True

6. False. Sample variability underestimates population variability.

7. False. The formula for sample variance divides by n - 1; the variance is 10.

8. True

9. True

10. True

Multiple-Choice Answers

1. b   2. a   3. a   4. a   5. b   6. b   7. b   8. d   9. a   10. a

Other Answers

1. SS = 32, n = 8, df = 7, s^2 = 4.57, and s = 2.14.

2. SS = 30, n = 5, σ^2 = 6, σ = 2.45

3. SS = 144, s^2 = 36, and s = 6

4. If a constant is added to each score, the mean also in increased by that constant. The deviation scores are not changed. If the deviation scores have not changed, then the squared deviations and SS will be unchanged. Thus, when a constant is added to every score in a distribution, the standard deviation is not changed.

Chapter 5

z-Scores

CHAPTER SUMMARY

Chapter 5 introduces the procedure for transforming scores (X values) into standardized z-scores. The process of changing X values into z-scores serves two purposes:

1. The z-score value specifies an exact location within a distribution.
2. Transforming X's to z-scores <u>standardizes</u> a distribution so that different distributions can be made comparable.

<u>z-Scores and Location</u>:   By itself, a raw score or X value provides very little information about how that particular score compares with other values in the distribution. A score of X = 53, for example, may be a relative low score, or an average score, or an extremely high score depending on the mean and standard

deviation for the distribution from which the score was obtained. If the raw score is transformed into a z-score, however, the value of the z-score tells exactly where the score is located relative to all the other scores in the distribution. The process of changing an X value into a z-score involves creating a signed number, called a z-score, such that

a. The sign of the z-score (+ or -) identifies whether the X value is located is above the mean (positive) or below the mean (negative).

b. The numerical value of the z-score corresponds to the number of standard deviations between X and the mean of the distribution.

Thus, a score that is located two standard deviations above the mean will have a z-score of +2.00. And, a z-score of +2.00 always indicates a location above the mean by two standard deviations.

This basic definition is usually sufficient to complete most z-score transformations. However, the definition can be written in mathematical notation to create a formula for computing z-scores.

$$z = \frac{X - \mu}{\sigma}$$

Also, the terms in the formula can be regrouped to create an equation for computing the value of X corresponding to any specific z-score.

$$X = \mu + z\sigma$$

In addition to knowing the basic definition of a z-score and the formula for a z-score, it is useful to be able to visualize z-scores as locations in a distribution (see Figure 5.2 in the text). Remember, $z = 0$ is in the center (at the mean), and the extreme tails correspond to z-scores of approximately -2.00 on the left and +2.00 on the right. Although more extreme z-score values are possible, most of the distribution is contained between $z = -2.00$ and $z = +2.00$.

The fact that z-scores identify exact locations within a distribution means that z-scores can be used as descriptive statistics and as inferential statistics. As descriptive statistics, z-scores describe exactly where each individual is located. As inferential statistics, z-scores determine whether a specific sample is representative of its population, or is extreme and unrepresentative. For example, a sample with a z-score near zero is a central, typical sample located near the population mean. On the other hand, a sample with a z-score value beyond 2.00 (or -2.00) would be considered an extreme or unusual sample, much different from the population mean.

z-Scores and Standardized Distributions: When an entire distribution of X values is transformed into z-scores, the resulting distribution of z-scores will always have a mean of zero and a standard deviation of one. The transformation does not change the shape of the original distribution and it does not change the location of any individual score relative to others in the distribution.

The advantage of standardizing distributions is that two (or more) different distributions can be made the same. For example, one distribution has $\mu = 100$ and $\sigma = 10$, and another distribution has $\mu = 40$ and $\sigma = 6$. When these distribution are transformed to z-scores, both will have $\mu = 0$ and $\sigma = 1$. Because z-score distributions all have the same mean and standard deviation, individual scores from different distributions can be directly compared. A z-score of $+1.00$ specifies the same location in all z-score distributions.

Although transforming X values into z-scores creates a standardized distribution, many people find z-scores burdensome because they consist of many decimal values and negative numbers. Therefore, it is often more convenient to standardize a distribution into numerical values that are simpler than z-scores. To create a simpler standardized distribution, you first select the mean and standard deviation that you would like for the new distribution. Then, z-scores are used to identify each individual's position in the original distribution and to compute the individual's position in the new distribution. Suppose, for example, that you want to standardize a distribution so that the new mean is $\mu = 50$ and the new standard deviation is $\sigma = 10$. An individual with a z-score of $z = -1.00$ in the original distribution would be assigned a score of $X = 40$ (below μ by one standard deviation) in the standardized distribution. Repeating this process for each individual score allows you to transform an entire distribution into a new, standardized distribution.

LEARNING OBJECTIVES

1. You should be able to describe and understand the purpose for z-scores.

2. You should be able to transform X values into z-scores or to transform z-scores into X values.

3. You should be able to describe the effects of standardizing a distribution by transforming the entire set of raw scores into z-scores.

4. Using z-scores, you should be able to transform any set of scores into a distribution with a predetermined mean and standard deviation.

NEW TERMS AND CONCEPTS

The following terms were introduced in this chapter. You should be able to define or describe each term and, where appropriate, describe how each term is related to other terms in the list.

raw score	An original, untransformed observation or measurement.
z-score	A standardized score with a sign that indicates direction from the mean (+ above μ and - below μ) and a numerical value equal to the distance from the mean measured in standard deviations.

z-score transformation	A transformation that changes raw scores (X values) into z-scores.
standard score	A score that has been transformed into a standard form.
standardized distribution	An entire distribution that has been transformed to create predetermined values for μ and σ.

NEW FORMULAS

$$z = \frac{X - \mu}{\sigma}$$

$$z\sigma = X - \mu = \text{deviation score}$$

$$X = \mu + z\sigma$$

STEP BY STEP

Changing X to z: The process of changing an X value to a z-score involves finding the precise location of X within its distribution. We will begin with a distribution with $\mu = 60$ and $\sigma = 12$. The goal is to find the z-score for $X = 75$.

Step 1: First determine whether X is above or below the mean. This will determine the sign of the z-score. For our example, X is above μ so the z-score will be positive.

Step 2: Next, find the distance between X and μ. For our example,
$$X - \mu = 75 - 60 = 15 \text{ points}$$
Note: Steps 1 and 2 simply determine a deviation score (sign and magnitude). If you are using the z-score formula, these two steps correspond to the numerator of the equation.

Step 3: Convert the distance from Step 2 into standard deviation units. In the z-score equation, this step corresponds to dividing by σ. For this example,
$$15/12 = 1.25$$

If you are using the z-score definition (rather than the formula), you simply compare the magnitude of the distance (Step 2) with the magnitude of the standard deviation. For this example, our distance of 15 points is equal to one standard deviation plus 3 more points. The extra 3 points are equal to one-quarter of a standard deviation, so the total distance is one and one-quarter standard deviations.

Step 4: Combine the sign from Step 1 with the number of standard deviations you obtained in Step 3. For this example,
$$z = +1.25$$

Changing z to X: The process of converting a z-score into an X value corresponds to finding the score that is located at a specified position in a distribution. Again, suppose we have a population with $\mu = 60$ and $\sigma = 12$. What is the X value corresponding to z $= -0.50$?

Step 1: The sign of the z-score tells whether X is above or below the mean. For this example, the X value we want is below μ.

Step 2: The magnitude of the z-score tells how many standard deviations there are between X and μ. For this example, the distance is one-half a standard deviation which is $(1/2)(12) = 6$ points.

Step 3:   Starting with the value of the mean, use the direction (Step 1) and the distance (Step 2) to determine the X value. For this example, we want to find the score that is 6 points below $\mu = 60$. Therefore,

$$X = 60 - 6 = 54$$

HINTS AND CAUTIONS

1. Rather than memorizing formulas for z-scores, we suggest that you rely on the definition of a z-score. Remember a z-score identifies a location by specifying the direction from the mean (+ or -) and the distance from the mean in terms of standard deviations.

2. When transforming scores from X to z (or from z to X) it is wise to check your answer by reversing the transformation. For example, given a population with $\mu = 54$ and $\sigma = 4$ a score of $X = 46$ corresponds to a z-score of

$$z = \frac{X - \mu}{\sigma} = \frac{46 - 54}{4} = \frac{-8}{4} = -2.00$$

To check this answer, convert the z-score back into an X value. In this case, $z = -2.00$ specifies a location below the mean by 2 standard deviations. This distance is

$$z\sigma = -2.00(4) = -8 \text{ points}$$

With a mean of $\mu = 54$, the score must be

$$X = 54 - 8 = 46.$$

=======

SELF-TEST

=======

True/False Questions

1. Any score with a value less than the population mean will have a negative z-score.

2. A positive z-score always indicates a position greater than the mean.

3. For a population with $\mu = 50$ and $\sigma = 10$, the z-score corresponding to $X = 45$ is $z = +0.50$.

4. For a population with $\mu = 50$ and $\sigma = 10$, the X value corresponding to $z = 1.50$ is $X = 65$.

5. One advantage of transforming X values into z-scores is that the transformation always creates a normal distribution.

6. For a population with $\mu = 30$, a score of $X = 22$ corresponds to $z = -2.00$. The standard deviation for the population is $\sigma = 4$.

7. For a population with σ = 10, a score of X = 48 corresponds to z = 2.00. The mean for the population is μ = 68.

8.   A population with μ = 100 and σ = 20 is transformed into z-scores. The resulting distribution of z-scores will have a mean of zero.

9.   A population with μ = 37 and σ = 6 is standardized to create a new distribution with μ = 100 and σ = 20. In this transformation a score of X = 40 from the original distribution will be transformed into a score of X = 110.

10.   Transforming an entire distribution of scores into z-scores will not change the shape of the distribution.

Multiple-Choice Questions

1. For a population with μ = 80 and σ = 12, the z-score corresponding to X = 74 is z = _____.
      a. 6
      b. -6
      c. .50
      d. -.50

2. For a population with μ = 100 and σ = 20, the X value corresponding to z = 1.50 is
      a.   X = 101.5
      b.   X = 115
      c.   X = 121.5
      d.   X = 130

3. A z-score of z = -2.00 indicates a position
    a. Below the mean by 2 points
    b. Below the mean by 2 times the standard deviation
    c. Above the mean by 2 points
    d. Above the mean by 2 times the standard deviation

4. For a population with σ = 10, a score that is located 20 points above the mean would have a z-score of _____.
    a. +20
    b. +2
    c. -2
    d. Cannot answer without knowing the mean

5. For a population with a mean of μ = 70, a score that is located 10 points below the mean would have a z-score of
    a. +1
    b. -1
    c. -10
    d. Cannot answer without knowing the standard deviation

6. In a population with σ = 8 a score of X = 42 corresponds to a z-score of z = -.50. What is the population mean?
    a. μ = 34
    b. μ = 38
    c. μ = 46
    d. μ = 48

7. If a population with $\mu = 60$ and $\sigma = 8$ is transformed into z-scores, then the resulting distribution of z-scores will have a mean of _____ and a standard deviation of _____.

    a. 0 and 1

    b. 60 and 1

    c. 0 and 8

    d. 60 and 8 (unchanged)

8. On an exam with $\mu = 52$ you have a score of $X = 48$. Which value for the standard deviation would give you a higher position in the class distribution.

    a. $\sigma = 1$

    b. $\sigma = 2$

    c. $\sigma = 4$

    d. Cannot determine from the information given

9. A population has $\mu = 50$. What value of σ would make $X = 55$ an extreme score in this population?

    a. $\sigma = 1$

    b. $\sigma = 5$

    c. $\sigma = 10$

    d. Cannot determine with the information given

10. You have a score of $X = 65$ on an exam. Which set of parameters would give you the best grade on the exam?

    a. $\mu = 60$ and $\sigma = 10$

    b. $\mu = 60$ and $\sigma = 5$

    c. $\mu = 70$ and $\sigma = 10$

    d. $\mu = 70$ and $\sigma = 5$

1. For a population with $\mu = 90$ and $\sigma = 25$ find the z-score corresponding to each of the following X values.
   a. X = 95
   b. X = 110
   c. X = 65
   d. X = 80

2. For a population with $\mu = 60$ and $\sigma = 6$ find the X value corresponding to each of the following z-scores.
   a. z = +1.50
   b. z = -0.50
   c. z = +2.00
   d. z = -1/3

3. On an exam with $\mu = 70$ and $\sigma = 10$, you have a score of X = 85.
   a. What is your z-score on this exam?
   b. If the instructor added 5 points to every score, what would happen to your z-score?
   c. If the instructor multiplied every score by 2, what would happen to your z-score?

4. A set of exam scores has $\mu = 48$ and $\sigma = 8$. The instructor would like to transform the scores into a standardized distribution with $\mu = 100$ and $\sigma = 20$. Find the transformed value for each of the following scores from the original population.
   a. X = 48
   b. X = 50
   c. X = 44
   d. X = 32

ANSWERS TO SELF-TEST

True/False Answers

1. True

2. True

3. False. The z-score is negative because X is below the mean.

4. True

5. False. Changing X to z does not change the shape of the distribution.

6. True

7. False. $X = 48$ is above the mean by 20 points so $\mu = 28$.

8. True

9. True

10. True

Multiple-Choice Answers

1. d   2. d   3. b   4. b   5. d   6. c   7. a   8. c   9. a   10. b

Other Answers

1. a. z = +0.20
   b. z = +0.80
   c. z = -1.00
   d. z = -0.40

2. a. X = 69   c. X = 72
   b. X = 57   d. X = 58

3. a. Your z-score is z = 1.50.
   b. Adding 5 points to every score would increase your score and the mean by 5 points. However, your z-score (your position within the distribution) would not change.
   c. Multiplying every score by 2 will multiply the mean, the standard deviation, and your score. However, your z-score (your position within the distribution) will not change.

4. a. X = 48 corresponds to z = 0. In the new distribution this location corresponds to X = 100.
   b. X = 50 corresponds to z = +0.25 which corresponds to X = 105 in the new distribution.
   c. X = 44 corresponds to z = -.50. X = 90 in the new distribution.
   d. X = 32 corresponds to z = -2.00. X = 60 in the new distribution.

Chapter 6

Probability

CHAPTER SUMMARY

In this chapter we introduce the concept of probability as a method for measuring and quantifying the likelihood of obtaining a specific sample from a specific population. We define probability as a fraction or a proportion. In particular, the probability of any specific outcome is determined by a ratio comparing the frequency of occurrence for that outcome relative to the total number of possible outcomes. Whenever the scores in a population are variable it is impossible to predict with perfect accuracy exactly which score or scores will be obtained when you take a sample from the population. In this situation researchers rely on probability to determine the relative likelihood for specific samples. Thus, although a researcher may not be able to predict exactly which value(s) will be obtained for a sample, it is possible to determine exactly which outcomes have high probability and which have low probability.

Probability is determined by a fraction or proportion. When a population of scores is represented by a frequency distribution, probabilities can be defined by proportions of the distribution. In graphs, probability can be defined as a proportion of area under the curve. For normal distributions, proportions (probabilities) can be found in the unit normal table. Because normal distributions are common, and because all normal distributions have the same shape (same proportions), it is possible to have one table that serves for all normal distributions, as long as it is standardized to z-scores. The table can be used to find the proportion associated with a specific score or to find the score associated with a specific proportion. In either case, it is necessary to use z-scores as an intermediate step to define precise locations within the distribution.

Probability is important because it establishes a link between samples and populations. For any known population it is possible to determine the probability of obtaining any specific sample. In later chapters we will use this link as the foundation for inferential statistics. The general goal of inferential statistics is to use the information from a sample to reach a general conclusion (inference) about an unknown population. Typically a researcher begins with a sample. If the sample has a high probability of being obtained from a specific population, then the researcher can conclude with some confidence that the sample actually came from that population. On the other hand, if the sample has a very low probability of being obtained from a specific population, then it is reasonable for the researcher to conclude that the specific population is probably not the source for the sample.

LEARNING OBJECTIVES

1. Know how to determine the probability of an event.

2. Be able to use the unit normal table to determine the probabilities for events that are normally distributed.

3. Be able to use the unit normal table to find the specific score associated with given probabilities or proportions.

4. Be able to find percentiles and percentile ranks for scores in a normal distribution.

NEW TERMS AND CONCEPTS

The following terms were introduced in this chapter. You should be able to define or describe each term and, where appropriate, describe how each term is related to other terms in the list.

probability

Probability is defined as a proportion, a specific part out of the whole set of possibilities.

proportion

A part of the whole usually expressed as a fraction.

random sample

A sample obtained using a process that gives every individual an equal chance of being selected and keeps the probability of being selected constant over a series of selections.

sampling with replacement

A sampling technique that returns the current selection to the population before the next selection is made. A required part of random sampling.

independent events

Two events are independent if the occurrence of either one has no effect on the probability that the other will occur.

normal distribution

A symmetrical, bell-shaped distribution with the greatest frequency in the center.

unit normal table

A table listing proportions corresponding to each z-score location in a normal distribution.

percentile	A score that is identified by the percentage of the distribution that falls below its value.
percentile rank	The percentage of a distribution that falls below a specific score.

NEW FORMULAS

$$p(A) = \frac{\text{Number of ways event A can occur}}{\text{Total number of possible outcomes}}$$

$$\text{Semi-Interquartile Range} = 0.67\sigma$$

STEP BY STEP

Finding the probability associated with a specified score. The general process involves converting the score (X) into a z-score, then using the unit normal table to find the probability associated with the z-score. To demonstrate this process, we will find the probability of randomly selecting a score greater than 95 from a normal distribution with $\mu = 100$ and $\sigma = 10$.

Step 1: Sketch the distribution and identify the mean and standard deviation. Then, find the approximate location of the specified score and draw a vertical line through the distribution. For this example, X = 95 is located below the mean by roughly one-half of the standard deviation.

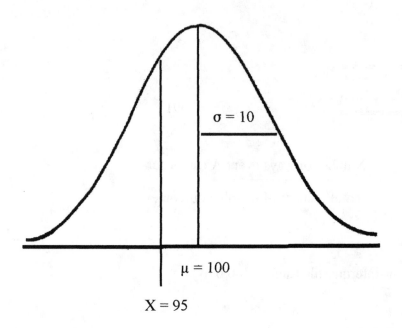

Step 2: Read the problem again to determine whether you want the proportion greater than the score (right of your line) or less than the score (left of the line). Then shade in the appropriate portion of the distribution. For this example we want the proportion consisting of scores greater than 95, so shade in the portion to the right of X = 95.

Step 3: Look at your sketch and make an estimate of the proportion that has been shaded. Remember, the mean divides the distribution in half with 50% on each side. For this example, we have shaded more than 50% of the distribution. The shaded area appears to be about 60% or 70% of the distribution.

Step 4: Transform the X value into a z-score. For this example, X = 95 corresponds to z = -0.50.

Step 5: Look up the z-score value in the unit normal table. (Ignore the + or - sign.) Find the two proportions in the table that are associated with your z-score and write these two proportions in the appropriate places on your figure. Remember, column B gives the proportion in the body of the distribution, and column C gives the area in the tail beyond z.

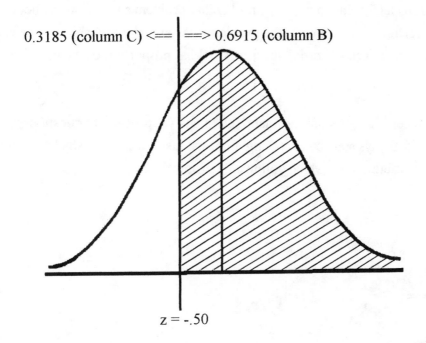

0.3185 (column C) <== | ==> 0.6915 (column B)

z = -.50

Step 6: You must indentify the column from the table that corresponds with the shaded area of your figure. For this example, the column B proportion, 0.6915, corresponds the shaded area.

Step 7: Compare your final answer with the estimate you made in Step 4. If your answer is not in agreement with your preliminary estimate, re-work the problem.

HINTS AND CAUTIONS

1. In using probability, you should be comfortable in converting fractions into decimals or percentages. These values all represent ways of expressing portions of the whole. If you have difficulty with fractions or decimals, review the section on proportions in the math review appendix of your textbook.

2. It usually helps to restate a probability problem as a question about proportion. For example, the problem, "What is the probability of selecting an ace from a deck of cards?" becomes, "What proportion of the deck is composed of aces?"

3. When using the unit normal table to answer probability questions, you should always start by sketching a normal distribution and shading in the area of the distribution for which you need a proportion.

SELF-TEST

True/False Questions

1. Probability values are always between 0 and 1.0.

2. Sampling with replacement is less important when the population size is small.

3. The definition of probability that is presented in this chapter requires random sampling.

4. The definition of a random sample implies sampling with replacement.

5. A container has 4 red marbles and 5 blue marbles. If one marble is selected randomly, then the probability of obtaining a red marble is 4/5 or 0.80.

6. A <u>random sample</u> of 2 marbles is obtained from a jar containing 10 red marbles and 20 blue marbles. If the first marble selected is red, then the probability that the second marble will be blue is $p = 20/29$.

7. The proportion in the tail beyond $z = +1.00$ is $p = .1587$ and the proportion in the tail beyond $z = -1.00$ is $p = -.1587$.

8. For a normal distribution, the proportion in the tail beyond $z = +1.00$ is exactly equal to the proportion in the tail beyond $z = -1.00$.

9. For any normal distribution, 84.13% of the individuals will have z-scores greater than $z = -1.00$.

10. For any normal distribution, the proportion in the tail beyond $z = +2.00$ is equal to 0.0228.

Multiple-Choice Questions

1. A jar contains 10 red marbles and 20 blue marbles. What is the probability of randomly selecting a red marble?
   a. 10/20
   b. 10/30
   c. 1/10
   d. 1/30

2. In a normal distribution, what z-score value separates the highest 20% of the scores from the rest of the distribution?

    a. $z = .84$

    b. $z = -.84$

    c. $z = 2.05$

    d. $z = .20$

3. In a normal distribution, what z-score value separates the lowest 10% of the scores from the rest of the distribution?

    a. $z = 1.28$

    b. $z = -1.28$

    c. $z = .25$

    d. $z = -.25$

4. For a normal distribution, what is the proportion in the tail beyond a z-score of $z = -1.50$?

    a. .9332

    b. -.9332

    c. .0668

    d. -.0668

5. If the tail on the right-hand side of a normal distribution contains exactly 2.5% of the scores, then what is the z-score value that separates the tail from the body of the distribution?

    a. $z = 2.50$

    b. $z = -2.50$

    c. $z = 1.96$

    d. $z = -1.96$

6. A normal distribution has $\mu = 80$ and $\sigma = 10$. What is the probability of randomly selecting a score greater than 85 from this distribution?

    a. $p = .50$
    b. $p = .25$
    c. $p = .3085$
    d. $p = .6915$

7. A normal distribution has $\mu = 100$ and $\sigma = 20$. What is the probability of randomly selecting a score less than 130 from this distribution?

    a. $p = .9032$
    b. $p = .9332$
    c. $p = .0968$
    d. $p = .0668$

8. A normal distribution has $\mu = 100$ and $\sigma = 20$. What score separates the top 40% from the rest of the distribution?

    a. $X = 110$
    b. $X = 105$
    c. $X = 90$
    d. $X = 95$

9. For any normal distribution, the 40th percentile corresponds to

    a. $z = 1.28$
    b. $z = -1.28$
    c. $z = 0.25$
    d. $z = -0.25$

10. For any normal distribution, the semi-interquartile range is

    a. 0.25σ
    b. 0.50σ
    c. 0.67σ
    d. 1.00σ

Other Questions

1. Assume a normal distribution for each question.
   a. What is the probability of obtaining a z-score greater than 1.25?
   b. What is the probability of obtaining a z-score less than 0.50?
   c. What proportion of the distribution consists of z-scores greater than -1.00?
   d. $p(z > 2.00) = $?
   e. $p(z < -.50) = $?

2. Find the z-score that separates a normal distribution into the following two portions:
   a. separate the lowest 75% from the highest 25%
   b. separate the lowest 90% from the highest 10%
   c. separate the lowest 35% from the highest 65%
   d. separate the lowest 42% from the highest 58%

3. Find the following probabilities for a normal distribution with $\mu = 80$ and $\sigma = 12$.
   a. $p(X > 86)$
   b. $p(X > 77)$
   c. $p(X < 95)$
   d. $p(X < 68)$

4. For a normal distribution with $\mu = 80$ and $\sigma = 12$, find the X value associated with each of the following proportions:
   a. What X value separates the distribution into the top 40% versus the bottom 60%?
   b. What is the minimum X value needed to be in the top 25% of the distribution?
   c. What X value separates the top 60% from the bottom 40% of the distribution?

True/False Answers

1. True

2. False. The requirement of replacement becomes more important when the sample size is small because the probabilities can change greatly without replacement.

3. True

4. True

5. False. The probability would be 4/9 because there are 9 marbles.

6. False. The probability would be 20/30. A random sample requires sampling with replacement

7. False. Both proportions are positive.

8. True

9. True

10. True

Multiple-Choice Answers

1. b   2. a   3. b   4. c   5. c   6. c   7. b   8. b   9. d   10. c

Other Answers

1 a. $p(z > 1.25) = 0.1056$
   b. $p(z < 0.50) = 0.6915$
   c. $p(z > -1.00) = 0.8413$
   d. $p(z > 2.00) = 0.0228$
   e. $p(z < -0.50) = 0.3085$

2. a. $z = 0.67$ c. $z = -0.39$
    b. $z = 1.28$ d. $z = -0.20$

3. a. $z = 0.50$ and $p = 0.3085$
    b. $z = -0.25$ and $p = 0.5987$
    c. $z = 1.25$ and $p = 0.8944$
    d. $z = -1.00$ and $p = 0.1587$

4. a. $z = 0.25$ and $X = 83$
    b. $z = 0.67$ and $X = 88.04$
    c. $z = -0.25$ and $X = 77$

Chapter 7

The Distribution
of Sample Means

In the previous two chapters we presented the statistical procedures for computing z-scores and finding probabilities associated with individual scores, X-values. In order to find z-scores or probabilities, the first requirement is that you must know about <u>all the possible X values</u>, that is, the entire distribution. A z-score tells where an individual X is located relative to all the other X values in the distribution. To find a probability, we simply identified a proportion of all the possible X values.

In Chapter 7 we extend the concepts of z-scores and probability to samples of more than one score. Specifically, we will compute z-scores and find probabilities for sample means. To accomplish this task, the first requirement is that you must know about <u>all the possible sample means</u>, that is, the entire distribution of $\overline{X}$'s. Once this distribution is identified, then

1. A z-score can be computed for each sample mean. The z-score tells where the specific sample mean is located relative to all the other sample means.

2. The probability associated with a specific sample mean can be defined as a proportion of all the possible sample means.

The Distribution of $\overline{X}$: The distribution of sample means is defined as the set of means from all the possible random samples of a specific size (n) selected from a specific population. This distribution has well-defined (and predictable) characteristics that are specified in the Central Limit Theorem:

1. The mean of the distribution of sample means is called the Expected Value of $\overline{X}$ and is always equal to the population mean μ.

2. The standard deviation of the distribution of sample means is called the Standard Error of $\overline{X}$ and is computed by

$$\sigma_{\overline{X}} = \frac{\sigma}{\sqrt{n}}$$

3. The shape of the distribution of sample means tends to be normal. It is guaranteed to be normal if a) the population from which the samples are obtained is normal, or b) the sample size is n = 30 or more.

Within this distribution, the location of each sample mean can be specified by a z-score,

$$z = \frac{\overline{X} - \mu}{\sigma_{\overline{X}}}$$

Because the distribution of sample means tends to be normal, the z-score values can be used with the unit normal table to obtain probabilities. The procedures for computing z-scores and finding probabilities for sample means are essentially the same as we used for individual scores (in Chapters 5 and 6). However, when you are using sample means, you must remember to consider the sample size (n) and compute the standard error ($\sigma_{\overline{X}}$) before you start any other computations. Also, you must be sure that the distribution of sample means satisfies at least one of the criteria for normal shape before you can use the unit normal table.

The concept of the distribution of sample means and its characteristics should be intuitively reasonable. First, you should realize that sample means are variable. If two (or more) samples are selected from the same population, the two samples probably will have different means. Second, although the samples will have different means, you should expect the sample means to be close to the population mean. That is, the sample means should "pile up" around μ. Thus, the distribution of sample means tends to form a normal shape with an expected value of μ. Finally, you should realize that an individual sample mean probably will not be identical to its population mean; that is, there will be some "error" between $\overline{X}$ and μ. Some sample means will be relatively close to μ and others will be relatively far away. The standard error provides a measure of the standard distance between $\overline{X}$ and μ.

The Standard Error of $\overline{X}$: Standard error is perhaps the single most important concept in inferential statistics. The standard error of $\overline{X}$ is defined as the standard deviation of the distribution of sample means and measures the standard distance between a sample mean and the population mean. Thus, the Standard Error of $\overline{X}$ provides a measure of how accurately a sample mean represents its corresponding population mean.

The magnitude of the standard error is determined by two factors: σ and n. The population standard deviation, σ, measures the standard distance between a single score (X) and the population mean. Thus, the standard deviation provides a measure of the "error" that is expected for the smallest possible sample, when n = 1. As the sample size is increased, it is reasonable to expect that the error should decrease. In simple terms, the larger the sample, the more accurately it should represent its population. The formula for standard error incorporates the intuitive relationship between standard deviation, sample size, and "error."

$$\sigma_{\overline{X}} = \frac{\sigma}{\sqrt{n}}$$

As the sample size increases, the error decreases. As the sample size decreases, the error increases. At the extreme, when n = 1, the error is equal to the standard deviation.

LEARNING OBJECTIVES

1. For any specific sampling situation, you should be able to define and describe the distribution of sample means by identifying its shape, the expected value of $\overline{X}$, and the standard error of $\overline{X}$.

2. You should be able to define and calculate the standard error of $\overline{X}$.

3. You should be able to compute a z-score that specifies the location of a particular sample mean within the distribution of sample means.

4. Using the distribution of sample means, you should be able to compute the probability of obtaining specific values for a sample mean obtained from a given population.

5. You should be able to incorporate a visual presentation of standard error into a graph presenting means for a set of different samples. In addition, you should be able to use the visual presentation of standard error to help determine whether the obtained difference between two sample means reflects a "real" difference in the populations, or whether the sample mean difference is simply due to chance.

The following terms were introduced in this chapter. You should be able to define or describe each term and, where appropriate, describe how each term is related to other terms in the list.

distribution of sample means · The set of sample means from all the possible random samples for a specific sample size (n) from a specific population

sampling distribution · A distribution of statistics (as opposed to a distribution of scores). The distribution of sample means is an example of a sampling distribution.

expected value of $\overline{X}$ · The mean of the distribution of sample means. The average of the $\overline{X}$ values.

standard error of $\overline{X}$ · The standard deviation of the distribution of sample means. The standard distance between a sample mean and the population mean.

the central limit theorem · A mathematical theorem that specifies the characteristics of the distribution of sample mean.

$$\sigma_{\bar{X}} = \frac{\sigma}{\sqrt{n}} \qquad \text{or} \qquad \sigma_{\bar{X}} = \sqrt{\frac{\sigma^2}{n}}$$

$$z = \frac{\bar{X} - \mu}{\sigma_{\bar{X}}}$$

STEP BY STEP

Computing Probabilities for Sample Means: You should recall that we have defined probability as being equivalent to proportion. Thus, the probability associated with a specific sample mean can be defined as a specific proportion of the distribution of sample means. Because the distribution of sample means tends to be normal, you can use z-scores and the unit normal table to determine proportions or probabilities. The following example demonstrates the details of this process.

For a normal population with $\mu = 60$ and $\sigma = 12$, what is the probability of selecting a random sample of $n = 36$ scores with a sample mean greater than 64? In symbols, $p(\bar{X} > 64) = ?$

Step 1: Rephrase the probability question as a proportion question. For this example, "Out of all the possible sample means for n = 36, what proportion have values greater than 64?"

Step 2:   We are looking for a specific proportion of "all the possible sample means." The set of "all possible sample means" is the distribution of sample means. Therefore, the next step is to sketch the distribution. Show the expected value and standard error in your sketch. Caution: Be sure to use the standard error, not the standard deviation.

For this example, the distribution of sample means will have an expected value of $\mu = 60$, a standard error of $\sigma_{\bar{x}} = 12/\sqrt{36} = 2$, and it will be a normal distribution because the original population is normal (also because n > 30).

Caution: If the distribution of sample means is not normal, you cannot use the unit normal table to find probabilities.

Step 3:   Find the approximate location of the specified sample mean and draw a vertical line through the distribution. For this example, $\bar{X} = 64$ is located above the mean by roughly two times the standard deviation.

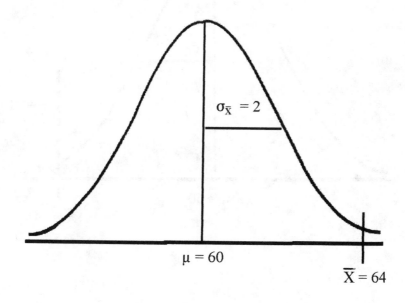

$\sigma_{\bar{x}} = 2$

$\mu = 60$

$\bar{X} = 64$

Step 4: Determine whether the problem asked for the proportion greater than or less than the specific $\overline{X}$. Then shade in the appropriate area in your sketch. For this example, we want the area greater than $\overline{X} = 64$ so shade in the area on the right-hand side of the line.

Step 5: Look at your sketch and make a preliminary estimate of the proportion that is shaded. For this example, we have shaded a very small part of the whole distribution, probably 5% or less.

Step 6: Compute the z-score for the specified sample mean. Be sure to use the z-score formula for sample means. For this example, $\overline{X} = 64$ corresponds to $z = +2.00$.

$$z = \frac{\overline{X} - \mu}{\sigma_{\overline{X}}} = \frac{64 - 60}{2} = \frac{4}{2} = 2.00$$

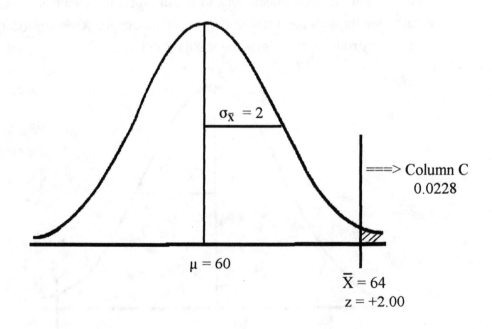

$\sigma_{\overline{X}} = 2$

===> Column C
0.0228

$\mu = 60$

$\overline{X} = 64$
$z = +2.00$

Step 7: Look up the z-score in the unit normal table and find the two proportions in columns B and C. For this example, the value in column C (the tail of the distribution) corresponds exactly to the proportion we want.

$$p(\overline{X} > 64) = p(z > +2.00) = .0228$$

Step 8: Compare your final answer with the preliminary estimate from Step 5. Be sure that your answer is in agreement with the common-sense estimate you made earlier.

HINTS AND CAUTIONS

1. Whenever you encounter a question about a sample mean you must remember to use the distribution of sample means and not the original population distribution. The population distribution contains scores (not sample means) and therefore should be used only when you have a question about an individual score ($n = 1$).

2. The key to working with the distribution of sample means is the standard error of $\overline{X}$:

$$\sigma_{\overline{X}} = \frac{\sigma}{\sqrt{n}}$$

Remember, larger samples tend to be more accurate (less error) than small samples. The sample size, n, is a crucial factor in determining the error between a sample and its population.

True/False Questions

1. The distribution of sample means is always a normal shaped distribution.

2. The distribution of sample mean is an example of a <u>sampling distribution</u>.

3. A population has $\mu = 40$ with $\sigma = 8$. The distribution of sample means for samples of size $n = 4$ selected from this population would have an expected value of 40.

4. If a sample consists of at least $n = 30$ scores, then the sample mean will be equal to the population mean.

5. A population has $\mu = 40$ with $\sigma = 8$. The distribution of sample means for samples of size $n = 4$ selected from this population would have a standard error of 10.

6. As sample size increases, the value of the standard error decreases.

7. It is impossible for the standard error to have a value larger than the standard deviation of the population from which the sample is selected.

8. A sample of $n = 9$ scores has a standard error of 10. Based on this information, the sample was obtained from a population with $\sigma = 30$.

9. A sample of $n = 4$ scores is selected from a population with $\mu = 30$ and $\sigma = 8$. If the sample mean is $\overline{X} = 34$, then this sample mean corresponds to a z-score of $z = .50$.

10. A sample is obtained from a population with $\sigma = 12$. If the sample mean has a standard error of 3, then the sample size is $n = 4$.

Multiple-Choice Questions

1. A population has $\mu = 80$ with $\sigma = 8$. The distribution of sample means for samples of size n = 4 selected from this population would have an expected value of _____.

    a. 8

    b. 80

    c. 40

    d. 20

2. A population has $\mu = 80$ with $\sigma = 8$. The distribution of sample means for samples of size n = 4 selected from this population would have a standard error of _____.

    a. 80

    b. 8

    c. 4

    d. 2

3. A sample of n = 25 scores is selected from a population with $\mu = 100$ with $\sigma = 20$. On average, how much error would be expected between the sample mean and the population mean?

    a. 25 points

    b. 20 points

    c. 4 points

    d. 0.8 points

4. A sample of n = 4 scores has a standard error of 12. What is the standard deviation of the population from which the sample was obtained?

    a. 48

    b. 24

    c. 6

    d. 3

5. A sample obtained from a population with σ = 10 has a standard error of 2 points. The sample size is
    a. n = 5
    b. n = 10
    c. n = 20
    d. n = 25

6. A sample of n = 16 scores is obtained from a population with μ = 70 and σ = 20. If the sample mean is $\overline{X}$ = 75, then the z-score corresponding to the sample mean is
    a. z = 0.25
    b. z = 0.50
    c. z = 1.00
    d. z = 2.00

7. As sample size increases, the standard error of $\overline{X}$
    a. increases
    b. decreases
    c. stays constant

8. Which of the following samples will have the smallest standard error
    a. A sample of n = 4 from a population with σ = 12
    b. A sample of n = 16 from a population with σ = 12
    c. A sample of n = 4 from a population with σ = 20
    d. A sample of n = 16 from a population with σ = 20

9. A sample is obtained from a population with μ = 50 and σ = 8. Which of the following samples would produce the most extreme z-scores (farthest from zero)?
    a. A sample of n = 4 scores with $\overline{X}$ = 52
    b. A sample of n = 16 scores with $\overline{X}$ = 52
    c. A sample of n = 4 scores with $\overline{X}$ = 54
    d. A sample of n = 16 scores with $\overline{X}$ = 54

10.   A sample is obtained from a population with $\mu = 100$ and $\sigma = 20$. Which of the following samples would produce the z-scores closest to zero?
   a.   A sample of $n = 25$ scores with $\overline{X} = 102$
   b.   A sample of $n = 100$ scores with $\overline{X} = 102$
   c.   A sample of $n = 25$ scores with $\overline{X} = 104$
   d.   A sample of $n = 100$ scores with $\overline{X} = 104$

Other Questions

1. A population has a mean of $\mu = 75$ and a standard deviation of $\sigma = 12$.
   a. If a single score is randomly selected from this population, how close on the average should the score be to the population mean?
   b. If a sample of $n = 4$ scores is randomly selected from this population, how close on the average should the sample mean be to the population mean?
   c. If a sample of $n = 36$ scores is randomly selected from this population, how close on the average should the sample mean be to the population mean?

2. Each of the following samples was obtained from a population with $\mu = 100$ and $\sigma = 10$. Find the z-score for each sample mean.
   a. $\overline{X} = 90$ for a sample of $n = 4$
   b. $\overline{X} = 90$ for a sample of $n = 25$
   c. $\overline{X} = 102$ for a sample of $n = 4$
   d. $\overline{X} = 102$ for a sample of $n = 100$

3. For a normal population with $\mu = 70$ and $\sigma = 9$
   a. What is the probability of obtaining a sample mean greater than 73 for a sample of $n = 36$ scores?
   b. What is the probability of obtaining a sample mean less than 73 for a sample of $n = 9$ scores?

4. Given a normal population with $\mu = 40$ and $\sigma = 4$, what is the probability of obtaining a sample mean between 39 and 41 for a sample of $n = 16$ scores?

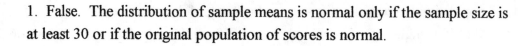

ANSWERS TO SELF-TEST

True/False Answers

1. False. The distribution of sample means is normal only if the sample size is at least 30 or if the original population of scores is normal.

2. True

3. True

4. False. The error gets smaller as the sample size increases, but you always expect some error.

5. False. The standard error is 4 points.

6. True

7. True

8. True

9. False. The standard error is 4 and $z = 1.00$.

10. False. The sample size is $n = 16$.

Multiple-Choice Answers

1. b   2. c   3. c   4. b   5. d   6. c   7. b   8. b   9. d   10. a

Other Answers

1. a. The standard deviation, $\sigma = 12$, measures the standard distance between a score and the population mean.
   b. For $n = 4$, the standard error is $12/\sqrt{4} = 6$.
   c. For a sample of $n = 36$. the standard error is 2 points.

2. a. The standard error is 5, and $z = -2.00$.
   b. The standard error is 2, and $z = -5.00$.
   c. The standard error is 5, and $z = +0.40$.
   d. The standard error is 1, and $z = +2.00$.

3. a. The standard error is 1.5. $p(\overline{X} > 73) = p(z > +2.00) = 0.0228$.
   b. The standard error is 3. $p(\overline{X} < 73) = p(z < 1.00) = .8413$.

4. The standard error is 1. The probability is $p(-1.00 < z < +1.00) = 0.6826$.
    [$1.00 - 2(.1587)$ That is, the whole distribution minus the two tails]

Chapter 8

Introduction to
Hypothesis Testing

CHAPTER SUMMARY

     Chapter 8 presents an introduction to the general procedure of hypothesis testing. Because hypothesis tests are probably the most commonly used inferential procedure, the basic concepts and terminology introduced in this chapter serve as a foundation for the remaining topics in this book. In very general terms, a hypothesis test begins with a population that has unknown parameters (usually, the population mean is unknown). A random sample is then obtained from the population, and the hypothesis test provides a standardized, formal procedure for using the sample data (typically, $\overline{X}$) as the basis for evaluating hypotheses about the population.

     Although hypothesis tests are used in a variety of situations, one common application is to help researchers determine whether or not a treatment has any effect on the individuals in the population. In this case, the researcher begins with a known population (before treatment). A random sample is selected and the treatment is administered to the sample. A hypothesis test is then used to make a comparison between the mean for the treated sample, and the mean for the original untreated population. If the sample statistic is substantially different from the original population parameter, we can conclude that the treatment has had an effect. However, if the sample does not look substantially different from the original population, we conclude that there does not seem to be any treatment effect.

Conclusions from hypothesis tests can get a bit tricky. Just because the sample mean (following treatment) is different from the original population mean does not necessarily indicate that the treatment has caused a change. You should recall (Chapter 7) that there usually is some discrepancy between a sample mean and a population mean simply as a result of sampling error. A hypothesis test is needed that to determine whether the obtained mean difference ($\overline{X}$ - μ) is simply due to sampling error or whether it really indicates a treatment effect.

The hypothesis-test procedure proceeds in four steps:

1. State the hypotheses and select an α level. The null hypothesis, H_0, always states the the treatment has no effect (no change, no difference). The α level establishes a criterion, or "cut-off", for making a decision about the hypotheses. The alpha level also determines the risk of a Type I error.

2. Locate the critical region. The critical region consists of outcomes that are very unlikely to occur if the null hypothesis is true. That is, the critical region is defined by sample means that are almost impossible to obtain just by chance. Their probability is $p < \alpha$. Thus, the alpha level is used to define precisely the terms "very unlikely" or "almost impossible."

3. Compute the test statistic. The test statistic (in this chapter a z-score) forms a ratio comparing the obtained difference between the sample mean and the hypothesized population mean versus the amount of difference we would expect just by chance (standard error).

4. A large value for the test statistic shows that the obtained mean difference is more than chance. If it is large enough to be in the critical region, we conclude that the difference is "significant" or that the treatment has a "significant effect." In this case we reject the null hypothesis. If the mean difference is not much larger than chance then the test statistic will have a low value. In this case, we conclude that the evidence from the sample is not sufficient, and the decision is fail to reject the null hypothesis.

LEARNING OBJECTIVES

1. Understand the logic of hypothesis testing.

2. Be able to state hypotheses and find the critical region.

3. Be able to assess sample data with a z-score and make a statistical decision about the hypotheses.

4. Know the difference between Type I and Type II errors.

5. When an experiment contains a prediction about the direction of a treatment effect, you should be able to incorporate the directional prediction into the hypothesis testing procedure and conduct a directional (one-tailed) hypothesis test.

The following terms were introduced in Chapter 8. You should be able to define or describe each term and, where appropriate, describe how each term is related to other terms in the list.

hypothesis testing	A statistical procedure that uses data from a sample to test a hypothesis about a population.
null hypothesis, H_0	The null hypothesis states that there is no effect, no difference, or no relationship.
alternative hypothesis, H_1	The alternative hypothesis states that there is an effect, there is a difference, or there is a relationship.
Type I error	A Type I error is rejecting a true null hypothesis. You have concluded that a treatment does have an effect when it actually does not.
Type II error	A Type II error is failing to reject a false null hypothesis. The test fails to detect a real treatment effect.
alpha (α)	Alpha is a probability value that is used to define the very unlikely outcomes if the null hypothesis is true. Alpha also is the probability of committing a Type I error.

level of significance	The level of significance is the alpha level, which measures the probability of a Type I error.
critical region	The critical region consist of outcomes that are very unlikely to be obtained if the null hypotheses is true. The term <u>very unlikely</u> is defined by α.
test statistic	A statistic that summarizes the sample data in a hypothesis test. The test statistic is used to determine whether or not the data are in the critical region.
beta (β)	Beta is the probability of a Type II error.
directional (one-tailed) test	A directional test is a hypothesis test that includes a directional prediction in the statement of the hypotheses and places the critical region entirely in one tail of the distribution.

NEW FORMULAS

$$P(\text{Type I Error}) = \alpha$$
$$P(\text{Type II Error}) = \beta$$

Using a Sample to Test a Hypothesis about a Population Mean: Although the hypothesis testing procedure is presented repeatedly in the textbook, we will demonstrate one more example here. As always, we will use the standard four-step procedure. The following generic example will be used for this demonstration.

The researcher begins with a known population, in this case a normal distribution with $\mu = 50$ and $\sigma = 10$. The researcher suspects that a particular treatment will produce a change in the scores for the individuals in the population. Because it is impossible to administer the treatment to the entire population, a sample of $n = 25$ individuals is selected and the treatment is given to the sample. After receiving the treatment, the average score for the sample is $\overline{X} = 53$. Although the experiment involves only a sample, the researcher would like to use the data to make a general conclusion about how the treatment affects the entire population.

Step 1: The first step is to state the hypotheses and select an alpha level. The hypotheses always concern an unknown population. For this example, the researcher does not know what would happen if the entire population were given the treatment. Nonetheless, it is possible to state hypotheses about the effect of the treatment. Specifically, the null hypothesis says that the treatment has no effect. According to H_0, the unknown population (after treatment) is identical to the original population (before treatment). In symbols,

H_0: $\mu = 50$ (After treatment, the mean is still 50)

The alternative to the null hypothesis is that the treatment does have an effect that causes a change in the population mean. In symbols,

H_1: $\mu \neq 50$ (After treatment, the mean is different from 50)

At this time you also select the alpha-level. Traditionally, α is set at .05 or .01. If there is particular concern about a Type I error, or if a researcher desires to present overwhelming evidence for a treatment effect, a smaller alpha-level can be used (such as $\alpha = .001$).

Step 2:   The next step is to locate the critical region. You should recall that the critical region is defined as the set of outcomes that are very unlikely to be obtained if the null hypothesis is true. We begin by looking at all the possible outcomes that could be obtained, then use the alpha level to determine the outcomes that are very unlikely. For this example, we look at the distribution of sample means for $n = 25$; that is, all the possible sample means that could be obtained if H_O were true.

The distribution of sample means will be normal because the original population is normal. The expected value is $\mu = 50$ (if H_0 is true), and the standard error for $n = 25$ is

$$\sigma_{\bar{x}} = \frac{\sigma}{\sqrt{n}} = \frac{10}{\sqrt{25}} = \frac{10}{5} = 2$$

With $\alpha = .05$, we want to identify the most unlikely 5% of this distribution. The boundaries for the extreme 5% are determined by z-scores of $z = \pm 1.96$.

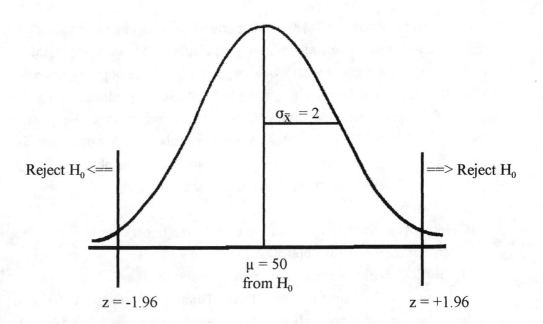

Step 3: Obtain the Sample Data and Compute the Test Statistic. For this example we obtained the sample mean of $\overline{X} = 53$. This sample mean corresponds to a z-score of,

$$z = \frac{\overline{X} - \mu}{\sigma_{\overline{X}}} = \frac{53 - 50}{2} = \frac{3}{2} = 1.50$$

Step 4: Make Your Decision. The z-score we obtained is not in the critical region. This means that our sample mean, $\overline{X} = 53$, is not an extreme or unusual value to be obtained from a population with $\mu = 50$. Therefore, we conclude that this sample does not provide sufficient evidence to conclude that the null hypothesis is wrong. Our statistical decision is to fail to reject H_0. The conclusion for the experiment is that the data do not indicate that the treatment has a significant effect. Note that the decision always consists of two parts:
     1) a statistical decision about the null hypothesis, and
     2) a conclusion about the outcome of the experiment.

HINTS AND CAUTIONS

1. When using samples with $n > 1$, we compute a z-score for the sample mean to determine if the sample data are unlikely. Be sure to use $\sigma_{\overline{X}}$ in the denominator, because the z-score is locating the sample mean within the distribution of sample means.

2. When stating the hypotheses for a directional test, remember that the predicted outcome (an increase or a decrease in μ) is stated in the alternative hypothesis (H_1).

True/False Questions

1. Rejecting the null hypothesis means that you have decided that the treatment does <u>not</u> have an effect.

2. In most research situations, the researcher hopes to reject the null hypothesis.

3. Committing a type I error means you failed to reject a false null hypothesis.

4. When committing a type I error, you are concluding that an effect exists, when in fact no change occurred.

5. A Type II error is most likely to occur when the treatment being evaluated has a very large effect.

6. Rejecting the null hypothesis with $\alpha = .05$ means that you have more confidence in your decision than if you have rejected the null hypothesis with $\alpha = .01$.

7. Changing α from .05 to .01 increases the risk of a Type I error.

8. If a sample mean falls into the critical region, then your decision should be to "fail to reject H_0."

9. As the alpha level increases, the boundaries for the critical region move farther out into the tails of the distribution.

10. If a researcher is predicting that a treatment will produce an increase in scores, then the critical region for a directional test would be located entirely in the left-hand tail of the distribution.

1. Which alpha level provides the smallest chance of committing a type I error?
    a. $\alpha = .01$
    b. $\alpha = .05$
    c. $\alpha = .10$
    d. $\alpha = .025$

2. For a regular two-tailed test with $\alpha = .01$, the boundaries for the critical region would be defined by z-scores of
    a. $z = \pm 1.96$
    b. $z = \pm 2.33$
    c. $z = \pm 2.58$
    d. $z = \pm 3.30$

3. In general, increasing the sample size (for example, from $n = 4$ to $n = 50$) will _____ the risk of a Type I error. (Assume α is held constant at .05)
    a. increase
    b. decrease
    c. have no influence on

4. If the sample data do not fall in the critical region, then the appropriate statement of the hypothesis test conclusion is
    a. Reject H_0
    b. Reject H_1
    c. Fail to reject H_0
    d. Fail to reject H_1

5. A Type I error involves
   a. Rejecting a false null hypothesis
   b. Rejecting a true null hypothesis
   c. Failing to reject a false null hypothesis
   d. Failing to reject a true null hypothesis

6. A hypothesis test is used to evaluate the effect of a treatment. If the test decision is to reject the null hypothesis with $\alpha = .05$, then
   a. If the experiment were repeated, then there is a 95% chance that the same result would be obtained
   b. If the experiment were repeated, then there is a 5% chance that the same result would be obtained
   c. There is a 5% chance that the treatment really has no effect
   d. There is a 95% chance that the treatment really has no effect

7. In a typical hypothesis testing situation, the null hypothesis makes a statement about
   a. The population before treatment
   b. The population after treatment
   c. The sample before treatment
   d. The sample after treatment

8. If a treatment has a very small effect, then a hypothesis test is likely to
   a. Result in a Type I error
   b. Result in a Type II error
   c. Correctly reject the null hypothesis
   d. Correctly fail to reject the null hypothesis

9. A researcher is conducting an experiment to evaluate a treatment that is expected to increase the scores for individuals in a population which is known to have a mean of $\mu = 80$. The results will be examined using a one-tailed hypothesis test. The correct statement of the null hypothesis is

    a. $\mu > 80$

    b. $\mu \geq 80$

    c. $\mu < 80$

    d. $\mu \leq 80$

10. A researcher is conducting a one-tailed test with $\alpha = .01$ to determine whether a treatment produces a significant increase in scores. What z-score value(s) would define the critical region for this test?

    a. beyond +2.33

    b. beyond -2.33

    c. beyond +2.58

    d. beyond -2.58

Other Questions

1. There always is some probability that a hypothesis test will lead to the wrong conclusion.

    a. Define a Type I error.

    b. Describe the consequences of a Type I error.

    c. What determines the probability of a Type I error.

    d. Define a Type II error.

    e. Describe the consequences of a Type II error.

2. Reaction times for a specific task are normally distributed with $\mu = 250$ and $\sigma = 50$. A particular sample of $n = 25$ subjects had a mean reaction time of $\overline{X} = 274$ milliseconds. Many of the subjects later complained about distracting noises during the test session.

    a. To assess whether their complaints are valid, determine if the sample data are significantly different from what would be expected. Use $\alpha = .05$.

    b. Would you have reached the same conclusion if alpha had been set at .01?

3. Scores for a standardized reading test are normally distributed with $\mu = 50$ and $\sigma = 6$ for sixth graders. A teacher suspects that his class is significantly above average for sixth grade and might need more challenging material. The class is given the standardized test, and the mean for the class of $n = 16$ students is $\overline{X} = 54.5$. Are these students significantly different from the typical sixth graders? Test with alpha set at .05.

ANSWERS TO SELF-TEST

True/False Answers

    1. False. Rejecting the null hypothesis indicates that the treatment did have an effect.

    2. True

    3. False. A Type I error means that you have rejected a true null hypothesis.

    4. True

5. False. Type II errors are common when the treatment effect is very small.

6. False. There is more risk of a Type I error when $\alpha = .05$.

7. False. A lower alpha level means less risk of a Type I error.

8. False. With data in the critical region, the decision is to reject the null hypothesis.

9. False. The critical region boundaries move farther out as α decreases.

10. False. With a predicted increase in scores, the critical region is in the right-hand tail.

Multiple-Choice Answers

1. a   2. c   3. c   4. c   5. b   6. c   7. b   8. b   9. d   10. a

Other Answers

1. a. A Type I error is rejecting a true null hypothesis.
   b. With a Type I error a researcher concludes that a treatment has an effect when in fact it does not. This can lead to a false report.
   c. The probability of a Type I error is the alpha level selected by the researcher.
   d. A Type II error is failing to reject a false null hypothesis.
   e. With a Type II error a researcher concludes that the data do not provide a convincing demonstration that the treatment has any effect when in fact it does. The researcher may choose to refine and repeat the experiment.

2. a. The critical region consists of values greater than $z = +1.96$ or less than $z = -1.96$. For this sample $z = 2.40$. Reject the null hypothesis and conclude that the data for this sample are significantly different from what would be expected by chance.

   b. With $\alpha = .01$, the critical region consists of values greater than $z = +2.58$ or less than -2.58. By this standard, the obtained z-score is not in the critical region, so we fail to reject H_0.

3. The null hypothesis states that these children are no different from the general population with $\mu = 50$. In symbols, $H_0: \mu = 50$. The critical region consists of z-scores beyond 1.96 or -1.96. The data produce a z-score of $z = 3.00$. Reject H_0 and conclude that this sample comes from a population with a mean that is different from $\mu = 50$.

Chapter 9

Introduction to
the t Statistic

CHAPTER SUMMARY

The t statistic introduced in Chapter 9 allows researchers to use sample data to test hypotheses about an unknown population mean. The particular advantage of the t statistic, compared to the z-score test in Chapter 8, is that the t statistic does not require any knowledge of the population standard deviation. Thus, the t statistic can be used to test hypotheses about a <u>completely unknown</u> population; that is, both μ and σ are unknown, and the only available information about the population comes from the sample. All that is required for a hypothesis test with t is a sample and a reasonable hypothesis about the population mean.

There are two general situations where this type of hypothesis test is used:

1. The t statistic is used when a researcher wants to determine whether or not a treatment causes a change in a population mean. In this case you must know the value of μ for the original, untreated population. A sample is obtained from the population and the treatment is administered to the sample. If the resulting sample mean is

significantly different from the original population mean, you can conclude that the treatment has a significant effect.

2. Occasionally a theory or other prediction will provide a hypothesized value for an unknown population mean. A sample is then obtained from the population and the t statistic is used to compare the actual sample mean with the hypothesized population mean. A significant difference indicates that the hypothesized value for μ should be rejected.

Estimating the Standard Error between $\overline{X}$ and μ: Whenever a sample is obtained from a population you expect to find some discrepancy or "error" between the sample mean and the population mean. This general phenomenon is known as "sampling error." The goal for a hypothesis test is to evaluate the significance of the observed discrepancy between a sample mean and the population mean. Specifically, the hypothesis test attempts to decide between the following two alternatives:

1. Is the discrepancy between $\overline{X}$ and μ within the margin of error? That is, is the difference between $\overline{X}$ and μ simply due to chance?

2. Is the discrepancy between $\overline{X}$ and μ more than would be expected by chance? That is, is the sample mean significantly different from the population mean?

The critical first step for the t statistic hypothesis test is to calculate exactly how much difference between $\overline{X}$ and μ is expected by chance. However, because the population standard deviation is unknown, it is impossible to compute the standard error of $\overline{X}$ as we did with z-scores in Chapter 8. Therefore, the t statistic requires that you use the sample data to compute an estimated standard error of $\overline{X}$. This calculation defines standard error exactly as it was defined in Chapters 7 and 8, but now we must use the sample variance, s^2, in place of the unknown population variance, σ^2 (or use sample standard deviation, s, in place of the unknown population standard deviation, σ). The resulting formula for estimated standard error is

$$ s_{\overline{X}} = \sqrt{\frac{s^2}{n}} \quad \text{or} \quad s_{\overline{X}} = \frac{s}{\sqrt{n}} $$

<u>Hypothesis Tests with the t Statistic</u>: The t statistic (like the z-score) forms a ratio. The top of the ratio contains the obtained difference between the sample mean and the hypothesized population mean. The bottom of the ratio is the standard error which measures how much difference is expected by chance.

$$t = \frac{\text{obtained difference}}{\text{standard error}} = \frac{\overline{X} - \mu}{s_{\overline{X}}}$$

A large value for t (a large ratio) indicates that the obtained difference between the data and the hypothesis is greater than would be expected by chance.

You should realize that the t statistic is very similar to the z-score used for hypothesis testing in Chapter 8. In fact, you can think of the t statistic as an "estimated z-score." The estimation comes from the fact that we are using the sample variance to estimate the unknown population variance. With a large sample, the estimation is very good and the t statistic will be very similar to a z-score. With small samples, however, the t statistic will provide a relatively poor estimate of z. The value of <u>degrees of freedom</u>, df = n - 1, is used to describe how well the t statistic represents a z-score. Also, the value of df will determine how well the distribution of t approximates a normal distribution. For large values of df, the t distribution will be nearly normal, but with small values for df, the t distribution will be flatter and more spread out than a normal distribution.

To evaluate the t statistic from a hypothesis test, you must select an α level, find the value of df for the t statistic, and consult the t distribution table. If the obtained t statistic is larger than the critical value from the table, you can reject the null hypothesis. In this case, you have demonstrated that the obtained difference between the data and the hypothesis (numerator of the ratio) is <u>significantly</u> larger than the difference that is expected by chance (the standard error in the denominator).

The hypothesis test with a t statistic follows the same four-step procedure that was used with z-score tests (Chapter 8):

1. State the hypotheses and select a value for α. (Note: The null hypothesis always states a specific value for μ.)

2. Locate the critical region.

(Note: You must find the value for df and use the t distribution table.)

3. Calculate the test statistic.

4. Make a decision (Either "reject" or "fail to reject" the null hypothesis).

LEARNING OBJECTIVES

1. Know when you must use the t statistic rather than a z-score for hypothesis testing.

2. Understand the concept of degrees of freedom and how it relates to the t distribution.

3. Be able to perform all of the necessary computations for hypothesis tests with the t statistic. This includes calculating the basic descriptive statistics for the sample (mean and variance) and the estimated standard error for $\overline{X}$.

NEW TERMS AND CONCEPTS

The following terms were introduced in this chapter. You should be able to define or describe each term and, where appropriate, describe how each term is related to other terms in the list.

t statistic	A statistic used to summarize sample data in situations where the population standard deviation is not known. The t statistic is similar to a z-score for a sample mean, but the t statistic uses an estimate of standard error.
estimated standard error	An estimate of the standard error that uses the sample variance (or standard deviation) in place of the corresponding population value.
degrees of freedom	Degrees of freedom = df = n - 1 measures the number of scores that are free to vary when computing SS for sample data. The value of df also describes how well a t statistic estimates a z-score.
t distribution	The distribution of t statistics is symmetrical and centered at zero like a normal distribution. A t distribution is flatter and more spread out than the normal distribution, but approaches a normal shape as df increases.

NEW FORMULAS

$$t = \frac{\overline{X} - \mu}{s_{\overline{X}}}$$

$$s_{\overline{X}} = \sqrt{\frac{s^2}{n}} \quad \text{or} \quad s_{\overline{X}} = \frac{s}{\sqrt{n}}$$

<u>Hypothesis Testing with the t Statistic</u>:   The t statistic presented in this chapter is used to test a hypothesis about an unknown population mean using the data from a single sample. Calculation of the t statistic requires the sample mean $\overline{X}$ and some measure of the sample variability, usually the sample variance, s^2. A hypothesis test with the t statistic uses the same four-step procedure that we use for all hypothesis tests. However, with a t statistic, you must compute the variance (or standard deviation) for the sample of scores and you must remember to use the t distribution table to locate the critical values for the test.   We will use the following example to demonstrate the t statistic hypothesis test.

A psychologist has prepared an "Optimism Test" that is administered yearly to graduating college seniors. The test measures how each graduating class feels about its future -- the higher the score, the more optimistic the class. Last year's class had a mean score of $\mu = 56$. A sample of $n = 25$ seniors from this year's class produced an average score of $\overline{X} = 59$ with SS = 2400. On the basis of this sample can the psychologist conclude that this year's class has a different level of optimism than last year's class? Test at the .05 level of significance.

Note that this test will use a t statistic because the population standard deviation is not known.

Step 1:   State Hypotheses and select an alpha level.
The statements of the null hypothesis and the alternative hypothesis are the same for the t statistic test as they were for the z-score test.

H$_0$: $\mu = 56$ (no change)
H$_1$: $\mu \neq 56$ (this year's mean is different)

For this example we are using $\alpha = .05$

Step 2: Locate the critical region. With a sample of $n = 25$, the t statistic will have $df = 24$.   For a two-tailed test with $\alpha = .05$ and $df = 24$, the critical t values are $t = \pm 2.064$.

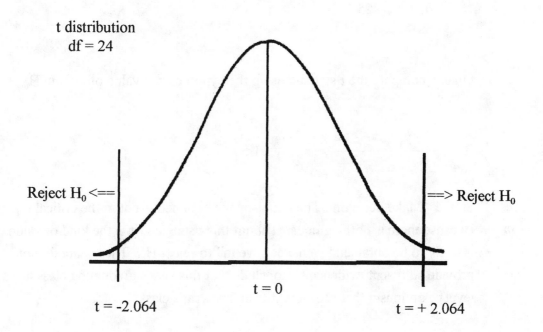

t distribution
df = 24

Reject H_0 <==

==> Reject H_0

t = 0

t = -2.064

t = + 2.064

Step 3: Obtain the data and compute the test statistic: For this sample we have $\overline{X} = 59$, $SS = 2400$, and $n = 25$. To compute the t statistic for these data, it is best to start by calculating the sample variance.

$$s^2 = \frac{SS}{n-1} = \frac{2400}{24} = 100$$

Next, use the sample variance to compute the estimated standard error. Remember, standard error provides a measure of the standard distance between a sample mean $\overline{X}$ and its population mean μ.

$$s_{\overline{X}} = \sqrt{\frac{s^2}{n}} = \sqrt{\frac{100}{25}} = \sqrt{4} = 2$$

Finally, compute the t statistic using the hypothesized value of μ from H_o.

$$t = \frac{\overline{X} - \mu}{s_{\overline{X}}} = \frac{59 - 56}{2} = \frac{3}{2} = 1.50$$

Step 4: Make decision. The t statistic we obtained is not in the critical region. Because there is nothing unusual about this t statistic (it is the kind of value that is likely to be obtained by chance), we fail to reject H_o. These data do not provide sufficient evidence to conclude that this year's graduating class has a level of optimism that is different from last year's class.

HINTS AND CAUTIONS

1. Students often confuse the formulas for sample variance (or standard deviation) and estimated standard error. The specific confusion is deciding when to divide by n and when to divide by n - 1.

Sample variance and standard deviation are <u>descriptive statistics</u> that were introduced in Chapter 4. You should recall that the sample mean describes the center of the distribution, and the sample standard deviation (or variance) describes how the scores are distributed around the mean. To compute the sample variance or the sample standard deviation, the correct denominator is df = n - 1.

The estimated standard error ($s_{\bar{x}}$) on the other hand, is primarily an inferential statistic that measures how accurately the sample mean represents its population mean. The degree of accuracy is largely determined by the size of the sample: The bigger the sample, the smaller the error. Thus, the magnitude of the standard error is determined by n. To compute the estimated standard error, the correct denominator is n.

2. When locating the critical region for a t test, be sure to consult the t distribution table, not the unit normal table.

SELF-TEST

True/False Questions

1. To compute a t statistic, you must use the sample variance (or standard deviation) to compute the estimated standard error for the sample mean.

2. If the sample variance increases, the estimated standard error will also increase.

3. A sample of n = 16 scores will produce a t statistic with df = 15.

4. A researcher reports a t statistic with df = 29. This t statistic was computed for a sample of n = 30 scores.

5. The distribution of t statistics tends to be flatter and more spread out than a normal distribution.

6. As the value of df increases, the t distribution tends to become flatter and more spread out.

7. As the sample variance increases, the value for the t statistic also increases. (Assume all other factors are held constant.)

8. As the sample size increases, the value of the estimated standard error tends to decrease. (Assume all other factors are held constant.)

9. For a two-tailed hypothesis test with $\alpha = .05$ using a sample of $n = 20$ scores, the critical values for t would be $t = \pm 2.086$.

10. For a one-tailed test with $\alpha = .01$ using a sample of $n = 16$, the critical value for t would be $t = 2.602$.

Multiple-Choice Questions

1. A sample of $n = 4$ scores has SS = 48. What is the variance for this sample?
   a. 12
   b. 16
   c. 4
   d. 2

2. A sample of $n = 4$ scores has SS = 48. What is the estimated standard error for this sample?
   a. 12
   b. 16
   c. 4
   d. 2

3. The estimated standard error, $s_{\bar{x}}$, provides a measure of
   a. How spread out the scores are in the sample
   b. How spread out the scores are in the population
   c. How much difference is reasonable to expect between the sample mean and the population mean
   d. How much difference is reasonable to expect between the t statistic and the corresponding z-score

4. What t values define the critical region for a regular two-tailed test using a sample of $n = 25$ scores and an alpha level of .05?
   a. $t = \pm1.711$
   b. $t = \pm1.708$
   c. $t = \pm2.064$
   d. $t = \pm2.060$

5. As the sample size increases, what happens to the critical values for t? Assume that the alpha level and all other factors remain constant.)
   a. The values increase
   b. The values decrease
   c. The values do not change when the sample size changes

6. A sample of $n = 15$ scores produces a t statistic of $t = -2.96$. If a researcher is using a regular two-tailed test with $\alpha = .01$, what decision should be made?
   a. Reject the null hypothesis
   b. Fail to reject the null hypothesis
   c. Cannot determine without more information

7. A sample of n = 25 scores has a mean of $\overline{X}$ = 46 and a variance of s^2 = 100. What is the estimated standard error for this sample?

    a. 10

    b. 5

    c. 4

    d. 2

8. A sample of n = 9 scores has a mean of $\overline{X}$ = 40 and a variance of s^2 = 9. If this sample is being used to test a null hypothesis stating that μ = 43, then what is the t statistic for the sample?

    a. t = -1/3

    b. t = -1.00

    c. t = -3.00

    d. t = -9.00

9. A researcher reports a t statistic with df = 20. Based on this information, how many individuals were in the sample?

    a. 19

    b. 20

    c. 21

    d. Cannot be determined without more information

10. A sample of n = 4 individuals is obtained from a population with μ = 80. Which set of sample statistics would produce the most extreme value for t?

    a. $\overline{X}$ = 84 and s^2 = 8

    b. $\overline{X}$ = 84 and s^2 = 32

    c. $\overline{X}$ = 88 and s^2 = 8

    d. $\overline{X}$ = 88 and s^2 = 32

1. A research study produces a t statistic of $t = 2.22$ for a sample of $n = 12$ scores.
   c. Is this t statistic sufficient to reject the null hypothesis using a two-tailed test with $\alpha = .05$?
   b. Is the t statistic sufficient to reject H_0 using a two-tailed test with $\alpha = .01$?

2. A researcher would like to evaluate the effect of a new cold medication on reaction time. It is known that under regular circumstances the distribution of reaction times is normal with $\mu = 200$. A sample of $n = 9$ subjects is obtained. Each person is given the new cold medication and 30 minutes later reaction time is measured for each individual. The average reaction time for this sample is $\overline{X} = 210$, and the sample has SS = 1800. Based on these data can the researcher conclude that the cold medication has a significant effect on reaction time. Test with $\alpha = .05$.
   a. Using symbols, state the hypotheses for this test.
   b. Locate the critical region for $\alpha = .05$.
   c. Calculate the t statistic for this sample.
   d. What decision should the researcher make?

3. A researcher obtains the following sample from a population with an unknown mean and unknown standard deviation.
   Sample: 9, 13, 9, 9
   a. Compute the mean, variance, and standard deviation for the sample.
   b. Use the sample variance to compute the estimated standard error, $s_{\overline{X}}$, for the sample mean.
   c. Use the sample to test the null hypothesis that the population mean is equal to 7. Use a two-tailed test with $\alpha = .05$. (Assume the population distribution is normal).

ANSWERS TO SELF-TEST

True/False Answers

1. True

2. True

3. True

4. True

5. True

6. False. As df increases, the t distribution becomes less spread out and more like a normal distribution.

7. False. As variance increases, the standard error also increases and the value of the t statistic moves toward zero.

8. True

9. False. With n = 20, df = 19 and the critical value is t = 2.093.

10. True

Multiple-Choice Answers

1. b 2. d 3. c 4. c 5. b 6. b 7. d 8. c 9. c 10. c

Other Answers

1. a. For $\alpha = .05$ and $df = 11$ the critical values are $t = \pm2.201$. The obtained
   value, $t = 2.22$, is in the critical region. Reject H_0.
   b. For $\alpha = .01$ and $df = 11$ the critical values are $t = \pm3.106$. The obtained t
   is not in the critical region. Fail to reject H_0.

2. a. H_0: $\mu = 200$ and H_1: $\mu \neq 200$
   b. With $df = 8$, the critical values are $t = \pm2.306$
   c. The sample variance is $s^2 = 225$, the standard error is 5, and $t = 2.00$.
   d. Fail to reject H_0. The data are not sufficient to indicate that the
   medication has a significant effect on reaction time.

3. a. $\overline{X} = 10$, $s^2 = 4$, and $s = 2$
   b. The estimated standard error is 1 point.
   c. If $\mu = 7$ (from H_0), then $t = 3.00$. This is not beyond the critical value of
   3.182 so the decision is to fail to reject the null hypothesis.

Chapter 10

Hypothesis Tests
with Two Independent Samples

CHAPTER SUMMARY

The independent-measures hypothesis test allows researchers to evaluate the difference between two population means using data from two separate samples, one sample representing each population. The identifying characteristic of this test is the existence of two separate or independent samples. Thus, an independent-measures design can be used to test for mean differences between two distinct populations (such as men versus women) or between two different treatment conditions (such as drug versus no-drug). The independent-measures design is used in situations where a researcher has no prior knowledge about either of the two populations (or treatments) being compared. In particular, the population means and standard deviations are all unknown. Because the population variances are not known, these values must be estimated from the sample data. Because we are trying to get the best estimate for population variance, we pool the two sample variances before computing the standard error. Then the resulting test statistic is the independent-measures t statistic.

As with all hypothesis tests, the general purpose of the independent-measures t test is to determine whether the sample mean difference obtained in a research study indicates a real mean difference between the two populations (or treatments) or whether the obtained difference is simply the result of sampling error. The hypothesis test provides a standardized, formal procedure for making this decision.

To prepare the data for analysis, the first step is to compute the sample mean and SS (or s, or s^2) for each of the two samples. The hypothesis test follows the same four-step procedure outlined in Chapters 8 and 9.

1. State the hypotheses and select an α level. For the independent-measures test, H_0 states that there is no difference between the two population means.

2. Locate the critical region. The critical values for the t statistic are obtained using degrees of freedom that are determined by adding together df for the first sample and df for the second sample.

3. Compute the test statistic. The t statistic for the independent-measures design has the same structure as the single sample t introduced in Chapter 9. However, in the independent-measures situation, all components of the t formula are doubled: there are two sample means, two population means, and two sources of error contributing to the standard error in the denominator.

4. Make a decision. If the t statistic ratio indicates that the obtained difference between sample means (numerator) is substantially greater than the difference expected by chance (denominator), we reject H_0 and conclude that there is a real mean difference between the two populations or treatments.

1. You should be able to describe and to recognize the experimental situations where an independent-measures t statistic is appropriate for statistical inference.

2. You should be able to use the independent-measures t statistic to test hypotheses about the mean difference between two populations (or between two treatment conditions).

3. You should be able to list the assumptions that must be satisfied before an independent-measures t statistic can be computed or interpreted.

NEW TERMS AND CONCEPTS

The following terms were introduced in Chapter 10. You should be able to define or describe each term and, where appropriate, describe how each term is related to others in the list.

independent-measures design

A research design that uses a separate sample for each treatment condition or each population being compared.

between-subjects design

An alternative term for an independent-measures design.

repeated-measures design	A research design that uses the same group of subject in all of the treatment conditions that are being compared.
within-subjects design	An alternative term for a repeated-measures design.
pooled variance	A single measure of sample variance that is obtained by averaging two sample variances. It is a weighted mean of the two variances.
homogeneity of variance	An assumption that the two populations from which the samples were obtained have equal variances.

NEW FORMULAS

$$t = \frac{(\overline{X}_1 - \overline{X}_2) - (\mu_1 - \mu_2)}{s_{\overline{X}_1 - \overline{X}_2}}$$

$$s_{\overline{X}_1 - \overline{X}_2} = \frac{s_P^2}{n_1} + \frac{s_P^2}{n_2}$$

$$s_P^2 = \frac{SS_1 + SS_2}{df_1 + df_2}$$

$$F_{max} = \frac{s^2 \text{ (largest)}}{s^2 \text{ (smallest)}}$$

STEP BY STEP

Hypothesis Tests with the Independent-Measures t Statistic. The independent-measures t statistic is used in situations where a researcher wants to test a hypothesis about the difference between two population means using the data from two separate (independent) samples. The test requires both sample means ($\overline{X}_1$ and $\overline{X}_2$), and some measure of the variability for each sample (usually SS). The following example will be used to demonstrate the independent-measures t hypothesis test.

A researcher wants to assess the damage to memory that is caused by chronic alcoholism. A sample of $n = 10$ alcoholics is obtained from a hospital treatment ward, and a control group of $n = 10$ non-drinkers is obtained from the hospital maintenance staff. Each person is given a brief memory test and the researcher records the memory score for each subject. The data are summarized as follows:

Alcoholics	Control
$\overline{X} = 43$	$\overline{X} = 57$
SS = 400	SS = 410

Step 1: State the hypotheses and select an alpha level. As always, the null hypothesis states that there is no effect.

H_0: $(\mu_1 - \mu_2) = 0$ (no difference)

The alternative hypothesis says that there is a difference between the two population means.

H_1: $(\mu_1 - \mu_2) \neq 0$

We will use $\alpha = .05$.

Step 2: Locate the critical region. With $n = 10$ in each sample, the t statistic will have degrees of freedom equal to,

$$df = n_1 + n_2 - 2 = 18$$

Sketch the entire distribution of t statistics with $df = 18$ and locate the extreme 5%. The critical values are $t = \pm 2.101$.

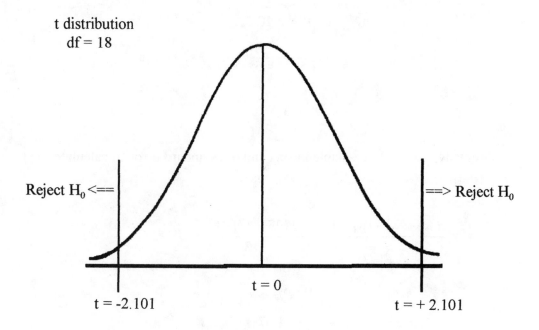

t distribution
df = 18

Reject H₀ <==

==> Reject H₀

t = 0

t = -2.101

t = + 2.101

Step 3: Compute the t statistic. It is easiest to begin by computing the pooled variance for the two samples.

$$s_P^2 = \frac{SS_1 + SS_2}{df_1 + df_2} = \frac{400 + 410}{9 + 9} = \frac{810}{18} = 45$$

Next, calculate the standard error for the t statistic.

$$s_{\bar{X}_1 - \bar{X}_2} = \sqrt{\frac{s^2_P}{n_1} + \frac{s^2_P}{n_2}} = \sqrt{\frac{45}{10} + \frac{45}{10}}$$

$$= \sqrt{4.5 + 4.5}$$

$$= \sqrt{9} = 3$$

Finally, use the two sample means and the standard error to calculate the t statistic.

$$t = \frac{(\bar{X}_1 - \bar{X}_2) - (\mu_1 - \mu_2)}{s_{\bar{X}_1 - \bar{X}_2}} = \frac{(43 - 57) - 0}{3}$$

$$= \frac{-14}{3} = -4.67$$

Step 4: Make decision. The t statistic for these data is in the critical region. This is a very unlikely outcome ($p < .05$) if H_0 is true, therefore, we reject H_0. The researcher concludes that there is a significant difference between the mean memory score for chronic alcoholics and the mean score for non-drinkers.

HINTS AND CAUTIONS

1. One of the most common errors in computing the independent-measures t statistic occurs when students confuse the formulas for pooled variance and

standard error. To compute the pooled variance, you combine the two samples into a single estimated variance. The formula for pooled variance uses a single fraction with SS in the numerator and df in the denominator:

$$s_P^2 = \frac{SS_1 + SS_2}{df_1 + df_2}$$

2. To compute the standard error, you add the separate errors for the two samples. In the formula for standard error these two separate sources of error appear as two separate fractions:

$$s_{\bar{x}_1 - \bar{x}_2} = \sqrt{\frac{s_P^2}{n_1} + \frac{s_P^2}{n_2}}$$

SELF-TEST

True/False Questions

1. The null hypothesis for an independent-measures study may be stated as, H_0: $\mu_1 - \mu_2 \neq 0$.

2. For the independent-measures t statistic, $df = n_1 + n_2 - 2$.

3. For an independent-measures study comparing two samples with $n = 8$ in each sample, the t statistic would have $df = 15$.

4. A researcher reports an independent-measures t statistic with $df = 20$. Based on the df value, you can conclude that the study used a total of 22 subjects.

5. One sample has n = 8 scores and SS = 40. A second sample has n = 4 scores and SS = 20. The pooled variance for these two samples is $s_p^2 = 60/10 = 6$.

6. The larger the difference between the two sample means, the larger the value for the independent-measures t statistic. (Assume all other factors are held constant.)

7. The larger the values for the two sample variances, the larger the value for the independent-measures t statistic. (Assume all other factors are held constant.)

8. You will be much more likely to detect a treatment effect when $s_{\bar{X}_1 - \bar{X}_2}$ is large.

9. There are two sources of error in the independent-measures study because there are two samples representing two populations.

10. The homogeneity assumption states that the two population means must be equal.

Multiple-Choice Questions

1. A researcher is comparing two treatment conditions with a sample of n = 5 in one treatment and a separate sample of n = 10 in the other. If the data are evaluated with an independent-measures t statistic, what is the df value for the statistic?
    a. 17
    b. 16
    c. 14
    d. 13

2. A researcher reports an independent measures t statistic with df = 18. What is the total number of subjects who participated in the research study?
    a. 16
    b. 17
    c. 19
    d. 20

3. One sample of n = 5 scores has SS = 36. A second sample of n = 7 scores has SS = 64. What is the value of the pooled variance (s_p^2) for these two samples?

    a. 100/12

    b. 100/10

    c. 36/5 + 64/7

    d. 36/4 + 64/6

4. A researcher computes the pooled variance for two samples and obtains a value of 120. If one of the sample has n = 5 scores and the second has n = 10 scores, then what is the value of the estimated standard error for the sample mean difference?

    a. $\sqrt{120/15}$

    b. $\sqrt{120/13}$

    c. $\sqrt{120/5 + 120/10}$

    d. $\sqrt{120/4 + 120/9}$

5. The null hypothesis for the independent-measures t test states

    a. $\mu_1 - \mu_2 = 0$

    b. $\overline{X}_1 - \overline{X}_2 = 0$

    c. $\mu_1 - \mu_2 \neq 0$

    d. $\overline{X}_1 - \overline{X}_2 \neq 0$

6. Which of the following sets of sample data would produce the largest value for an independent-measures t statistic. Assume than n = 10 for all samples. Note: You should not need to do any serious calculations to answer this question.

    a. First sample: $\overline{X}$ = 30 and SS = 10. Second Sample: $\overline{X}$ = 35 and SS = 10

    b. First sample: $\overline{X}$ = 30 and SS = 10. Second Sample: $\overline{X}$ = 50 and SS = 10

    b. First sample: $\overline{X}$ = 30 and SS = 50. Second Sample: $\overline{X}$ = 35 and SS = 50

    d. First sample: $\overline{X}$ = 30 and SS = 50. Second Sample: $\overline{X}$ = 50 and SS = 50

7. In an independent-measures t hypothesis test, how does the magnitude for sample variance influence the size of the t statistic?
   a. The larger the sample variance, the larger the t value (farther from zero)
   b. The larger the sample variance, the smaller the t value (closer to zero)
   c. The magnitude of the sample variance has no influence on the t value.

8. In an independent-measures t hypothesis test, how is the t statistic related to the amount of difference between the two sample means?
   a. The larger the difference between means, the larger the value for t (farther from zero)
   b. The larger the difference between means, the smaller the value for t (closer to zero)
   c. The magnitude of the t statistic is not related to the mean difference between samples

9. One sample has n = 25 scores and a second sample has n = 15 scores. If the pooled variance is computed for the two samples, then the value for the pooled variance will be
   a. Half-way between the two sample variances
   b. Between the two sample variances, but closer to the variance of the larger sample
   c. Between the two sample variances, but closer to the variance of the smaller sample
   d. Larger than either of the two sample variances.

10. Which of the following research situations would be most likely to use an independent-measures design?
   a. Examine the development of vocabulary as a group of children mature from age 2 to age 3
   b. Examine the long-term effectiveness of a stop-smoking treatment by interviewing subjects 2 months and 6 months after the treatment ends
   c. Compare the mathematics skills for 9th grade boys versus 9th grade girls
   d. Compare the blood-pressure readings before medication and after medication for a group a patients with high blood pressure

Other Questions

1. A researcher would like to demonstrate how different schedules of reinforcement can influence behavior. Two separate groups of rats are trained to press a bar in order to receive a food pellet. One group is trained using a fixed ratio schedule where they receive one pellet for every 10 presses of the bar. The second group is trained using a fixed interval schedule where they receive one pellet for the first bar press that occurs within a 30 second interval. Note that the second group must wait 30 seconds before another pellet is possible no matter how many times the bar is pressed. After 4 days of training, the researcher records the response rate (number of presses per minute) for each rat. The results are summarized as follows:

Fixed Ratio	Fixed Interval
$n = 4$	$n = 8$
$\bar{X} = 30$	$\bar{X} = 18$
$SS = 90$	$SS = 150$

Do these data indicate that there is a significant difference in responding for these two reinforcement schedules? Test at the .05 level of significance.

2. A psychologist is examining the educational advantages of a preschool program. A sample of 24 fourth grade children is obtained. Half of these children had attended preschool and the others had not. The psychologist records the scholastic achievement score for each child and obtained the following data:

Preschool	No Preschool
8 6 8	8 5 7
9 7 7	6 8 5
6 9 8	7 5 6
9 7 8	7 5 6

Do these data indicate that participation in a preschool program gives children a significant advantage in scholastic achievement? Use a one-tailed test at the .05 level.

3. The following data are from two separate samples. Does it appear that these two samples came from the same population or from two different populations?

    a. Use an F-max test to determine whether there is evidence for a significant difference between the two population variances. Use the .05 level of significance.

    b. Use an independent-measures t test to determine whether there is evidence for a significant difference between the two population means. Again, use $\alpha = .05$.

Sample 1	Sample 2
$n = 10$	$n = 10$
$\bar{X} = 32$	$\bar{X} = 18$
$SS = 890$	$SS = 550$

ANSWERS TO SELF-TEST

True/False Answers

1. False. The null hypothesis states that the mean difference is zero.

2. True

3. False. $df = (n_1 - 1) + (n_2 - 1) = 7 + 7 = 14$

4. True

5. True

6. True

7. False. Increases in variance produce a smaller value for t (nearer to zero).

8. False. Larger error obscures any treatment effect.

9. True

10. False. The homogeneity assumptions states that the population variances are equal.

Multiple-Choice Answers

1. d   2. d   3. b   4. c   5. a   6. b   7. b   8. a   9. b   10. c

1. With df $= 10$, the critical t values are t $= \pm2.228$. These data have a pooled variance of 24 and produce a t statistic of t $= 4.00$. Reject H_0. The data provide sufficient evidence to conclude that there is a significant difference between the two schedules.

2. The psychologist expects the preschool children (sample 1) to have higher scores, so the hypotheses are,

H_0: $(\mu_1 - \mu_2) \leq 0$ (preschool is not higher)

H_1: $(\mu_1 - \mu_2) > 0$ (preschool is higher)

The critical region consists of t values greater than t $= +1.717$. For these data $\overline{X}_1 = 7.67$ and $\overline{X}_2 = 6.25$, $SS_1 = 12.67$ and $SS_2 = 14.25$. The pooled variance is 1.22, and t $= 3.16$. Reject H_0 and conclude that the preschool children score significantly higher on the scholastic achievement test.

3. a. For these data, F-max $= 1.62$. The critical value for $\alpha = .05$ is 4.03. Fail to reject H_0. There is insufficient evidence to conclude that the two population variances are different.

   b. For these data the pooled variance is 80 and the t statistic is t $= 14/4 = 3.50$. Reject H_0 and conclude that the two population means are different.

Chapter 11

Hypothesis Tests
With Related Samples

In a <u>repeated-measures design,</u> a single sample of individuals is obtained and each individual is measured in both of the treatment conditions being compared. Thus, the data consist of two scores for each individual. In a similar design, called a <u>matched-subjects design,</u> each individual in one treatment condition is matched one-to-one with a corresponding individual in the second treatment. The matching is accomplished by selecting pairs of subjects so that the two subjects in each pair have identical (or nearly identical) scores on the variable that is being used for matching. Thus, the data consist of pairs of scores with each pair corresponding to a matched set of two "identical" subjects.

<u>Hypothesis Tests with the Repeated-Measures t:</u> The repeated-measures t statistic allows researchers to test hypothesis about the population mean difference between

two treatment conditions using sample data from either a repeated-measures or a matched-subjects research study. The key element with either type of research design is that the sample of individuals in one treatment condition is matched exactly, one-to-one, with the sample of individuals in the second treatment condition. Thus, the scores in the first treatment are statistically related to the scores in the second treatment. In this situation it is possible to compute a <u>difference score</u> for each individual (or each matched pair):

$$\text{difference score} = D = X_2 - X_1$$

The sample of difference scores is then used to test hypotheses about the population of difference scores. The null hypothesis states that the population of difference scores has a mean of zero,

$$H_0: \mu_D = 0$$

In words, the null hypothesis says that there is no consistent or systematic difference between the two treatment conditions. According to the null hypothesis, any non-zero mean difference that is observed for a sample is simply the result of sampling error. (Note: The null hypothesis does not claim that each individual subject will have a zero difference between treatments. Some subjects will show a positive change from one treatment to the other, and some subjects will show a negative change. On average, however, the entire population will show a mean difference of zero.)

The alternative hypothesis states that there is a real, non-zero difference between the treatments:

$$H_1: \mu_D \neq 0$$

According to the alternative hypothesis, the sample mean difference obtained in the research study is a reflection of the true mean difference that exists in the population.

The repeated-measures t statistic forms a ratio with exactly the same structure as the single-sample t statistic presented in Chapter 9. The numerator of the t statistic measures the difference between the sample mean and the hypothesized population mean. The bottom of the ratio is the standard error, which measures how much difference is expected by chance.

$$t = \frac{\text{obtained difference}}{\text{standard error}} = \frac{\bar{D} - \mu_D}{s_{\bar{D}}}$$

For the repeated-measures t statistic, all calculations are done with the sample of difference scores. The mean for the sample appears in the numerator of the t statistic and the variance of the difference scores is used to compute the standard error in the denominator. As usual, the standard error is computed by

$$s_{\bar{D}} = \sqrt{\frac{s^2}{n}} \quad \text{or} \quad s_{\bar{D}} = \frac{s}{\sqrt{n}}$$

Variance and Individual Differences: The first step in the calculation of the repeated-measures t statistic is to find the difference score for each subject. This simple process has two very important consequences.

1. First, the D score for each subject provides an indication of how much difference there is between the two treatments. If all of the subjects show roughly the same D scores, then you can conclude that there appears to be a consistent, systematic difference between the two treatments. You should also note that when all the D scores are similar, the variance of the D scores will be small, which means that the standard error will be small and the t statistic is more likely to be significant.

2. Also, you should note that the process of subtracting to obtain the D scores removes the individual differences from the data. That is, the initial differences in performance from one subject to another are eliminated. Removing individual differences also tends to reduce the variance, which creates a smaller standard error and increases the likelihood of a significant t statistic.

The following data demonstrate these points:

Subject	X_1	X_2	D
A	9	16	7
B	25	28	3
C	31	36	5
D	58	61	3
E	72	79	7

First, notice that all of the subjects show an increase of roughly 5 points when they move from treatment 1 to treatment 2. Because the treatment difference is very consistent, the D scores are all clustered close together will produce a very small value for s^2. This means that the standard error in the bottom of the t statistic will be very small.

Second, notice that the original data show big differences from one subject to another. For example, subject B has scores in the 20's and subject E has scores in the 70's. However, these big individual differences are eliminated when the difference scores are calculated. Because the individual differences are removed, the D scores are usually much less variable that the original scores. Again, a smaller variance will produce a smaller standard error, which will increase the likelihood of a significant t statistic.

═══════════════

LEARNING OBJECTIVES
═══════════════

1. Know the difference between independent-measures and related-samples experimental designs.

2. Know the difference between a repeated-measures and a matched-subjects experimental design.

3. Be able to perform the computations for the related-samples t test.

4. Understand the advantages and disadvantages of the repeated-measures design and when this type of study is appropriate.

NEW TERMS AND CONCEPTS

The following terms were introduced in this chapter. You should be able to define or describe each term and, where appropriate, describe how each term is related to other terms in the list.

repeated-measures design	A research study where the same sample of individuals is measured in all of the treatment conditions.
matched-samples design	A research study where the individuals in one sample are matched one-to-one with the individuals in a second sample. The matching is based on a variable considered relevant to the study.
related-samples t statistic	The single sample t statistic applied to a sample of difference scores (D values) and the corresponding population of difference scores.
difference scores	The difference between two measurements obtained for a single subject. $D = X_2 - X_1$

estimated standard error of $\bar{D}$	An estimate of the standard distance between a sample mean difference $\bar{D}$ and the population mean difference μ_D
individual differences	The naturally occurring differences from one individual to another that may cause the individuals to have different scores.
carry-over effects	The aftereffects of one treatment that may influence the scores in the following treatment condition.

NEW FORMULAS

$$D = X_2 - X_1$$

$$t = \frac{\bar{D} - \mu_D}{s_{\bar{D}}}$$

$$s_{\bar{D}} = \sqrt{\frac{s^2}{n}} \quad \text{or} \quad s_{\bar{D}} = \frac{s}{\sqrt{n}}$$

Hypothesis Testing with the Related-Samples t. The related-samples t statistic is used to test for a mean difference (μ_D) between two treatment conditions using data from a single sample of subjects where each individual is measured first in one treatment condition and then in the second condition. This test statistic also is used for matched-subjects designs which consists of two samples with the subjects in one sample matched one-to-one with the subjects in the second sample. Often, a repeated-measures experiment consists of a "before/after" design where each subject is measured before treatment and then again after treatment. The following example will be used to demonstrate the related-samples t test.

A researcher would like to determine whether a particular treatment has an effect on performance scores. A sample of n = 16 subjects is selected. Each subject is measured before receiving the treatment and again after treatment. The researcher records the difference between the two scores for each subject. The difference scores averaged $\overline{D} = -6$ with SS = 960.

Step 1: State the hypotheses and select an alpha level. In the experiment the treatment was given to a sample, but the researcher wants to determine whether the treatment has any effect for the general population. As always, the null hypothesis says that there is no effect.

H_0: $\mu_D = 0$ (on average, the before/after difference for the population is zero)

The alternative hypothesis states that the treatment does produce a difference.

H_1: $\mu_D \neq 0$

We will use $\alpha = .05$.

Step 2: Locate the critical region. With a sample of n = 16, the related-samples t statistic will have df = 15. Sketch the distribution of t statistics with df = 15 and locate the extreme 5% of the distribution. The critical boundaries are t = ±2.131.

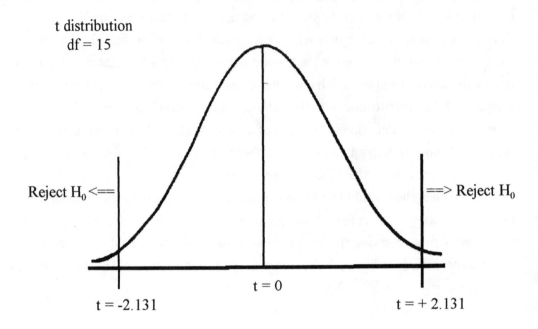

t distribution
df = 15

Reject H_0 <==

==> Reject H_0

t = 0

t = -2.131

t = + 2.131

Step 3: Calculate the test statistic. As with all t statistics, it is easier to begin the calculation with the denominator of the t formula. For this example, the variance for the difference scores is

$$s^2 = \frac{SS}{n-1} = \frac{960}{15} = 64$$

With a sample of n = 16, the standard error is

$$s_{\bar{D}} = \sqrt{\frac{s^2}{n}} = \sqrt{\frac{64}{16}} = \frac{8}{4} = 2$$

Finally, substitute the sample mean difference and the standard error in the t formula,

$$t = \frac{\overline{D} - \mu_D}{s_{\overline{D}}} = \frac{-6 - 0}{2} = \frac{-6}{2} = -3.00$$

Step 4: Make Decision. The t statistic is in the critical region. This is a very unlikely value for t if H_0 is true. Therefore, we reject H_0. The researcher concludes that the treatment does have a significant effect on performance scores.

HINTS AND CAUTIONS

1. It is important to remember that the related-samples analysis is based on difference scores (D scores). Therefore, the computations of s^2 and $s_{\overline{D}}$ are based on the sample of D scores.

2. When calculating the difference scores, be sure all of them are obtained by subtracting in the same direction. That is, you may either subtract 1st - 2nd, or 2nd - 1st as long as the same method is used throughout.

True/False Questions

1. A researcher would like to compare two treatment conditions using a sample of 30 scores in each treatment. If a repeated-measures design is used, the study will require a total of n = 60 subjects.

2. A researcher compares two treatment conditions using a repeated measures design with a sample of n = 20 subjects. The t statistic for this study will have df = 19.

3. A researcher reports a t statistic with df = 18 for a repeated-measures design. This research study used a total of n = 20 subjects.

4. The original data for a repeated-measures t statistic consist of two scores for each subject.

5. One concern in the evaluation of research results is that the subjects in one treatment condition may be substantially different (older, smarter, and so on) than the subjects in another condition. However, this is not a problem with a repeated-measures design.

6. After the difference scores (D values) have been computed, you do not need the original scores to complete the calculations for the repeated-measures t statistic.

7. As the variance of the difference scores increases, the value of the t statistic decreases (closer to zero).

8. As the size of the difference scores increases, the value of the t statistic decreases (moves closer to zero).

9. A repeated-measures test usually is more likely to detect a real treatment effect than an independent-measures test because the repeated-measures design reduces the standard error by removing individual differences.

10. One concern with a repeated-measures design is that the results may be distorted by carry-over effects.

Multiple Choice Questions

1. A researcher plans to conduct a research study comparing two treatment conditions with a total of 20 scores in each treatment. Which of the following designs would require the smallest number of subjects?
    a. An independent-measures design
    b. A repeated-measures design
    c. A matched-subjects design
    d. All of the above would require the same number of subjects

2. The following data were obtained from a repeated-measures research study. What is the value of $\overline{D}$ for these data?

		Subject	1st	2nd
a.	4	#1	8	10
b.	4.6	#2	6	12
c.	5	#3	10	7
d.	20	#4	9	17
		#5	7	14

3. A researcher reports a t statistic with df = 24 from a repeated-measures research study. How many subjects participated in the study?
    a. n = 11
    b. n = 13
    c. n = 23
    d. n = 25

4. The following data were obtained from a repeated-measures research study. What is the value of df for the t statistic?

a. df = 4
b. df = 5
c. df = 8
d. df = 9

Subject	1st	2nd
#1	8	10
#2	6	12
#3	10	7
#4	9	17
#5	7	14

5. A researcher uses a repeated-measures study to compare two treatment conditions with a set of 20 scores in each treatment. What would be the value of df for the repeated-measures t statistic?

a. df = 18
b. df = 19
c. df = 38
d. df = 39

6. In general, if the magnitude of the variance of the difference scores increases, then the value of the t statistic will

a. increase (move farther toward the tail of the distribution)
b. decrease (move toward 0 at the center of the distribution)
c. stay the same - the t statistic is not affected by the variance of the difference scores.

7. Which of the following samples will produce the largest value for a t statistic? Assume each sample has n = 10 scores.

a. $\bar{D} = 5$ with SS = 20
b. $\bar{D} = 10$ with SS = 20
c. $\bar{D} = 5$ with SS = 40
d. $\bar{D} = 10$ with SS = 40

8. The null hypothesis for a repeated-measures test states
   a. Each individual will have a difference score of $D = 0$
   b. The overall sample will have a mean difference of $\overline{D} = 0$
   c. The entire population will have a mean difference of $\mu_D = 0$
   d. All of the above
   e. None of the above

9. The data from a repeated-measures research study show that all of the subjects scored about 10 points higher in treatment I than in treatment II. These data will produce
   a. a small sample variance and a large t statistic
   b. a small sample variance and a small t statistic
   c. A large sample variance and a large t statistic
   d. A large sample variance and a small t statistic

10. The data from a repeated-measures research study show that the difference between treatment I and treatment II is large and positive for some subjects, large and negative for some subjects, and near zero for some subjects. These data will produce
   a. a small sample variance and a large t statistic
   b. a small sample variance and a small t statistic
   c. A large sample variance and a large t statistic
   d. A large sample variance and a small t statistic

Other Questions

1. A researcher would like to conduct a study comparing two treatment conditions with 30 individuals measured in each treatment.
   a. How many subjects would be needed if the researcher uses an independent-measures design?
   b. How many subjects would be needed if the researcher uses a repeated-measures design?

c. How many subjects would be needed if the researcher uses a matched-subjects design?

2. Calculate the difference scores and the mean difference, $\bar{D}$, for the following sample.

Subject	1st	2nd	D
#1	8	10	
#2	6	12	
#3	10	7	
#4	9	17	
#5	7	14	

3. A researcher would like to test the effect of a new diet drug on the activity level of animals. A sample of $n = 16$ rats is obtained and each rat's activity level is measured on an exercise wheel for one hour prior to receiving the drug. Thirty minutes after receiving the drug, each rat is again tested on the activity wheel. The data show that the rats increased their activity by an average of $\bar{D} = 21$ revolutions with $SS = 6000$ after receiving the drug. Do these data indicate that the drug had a significant effect on activity. Test at the .05 level.

a. Using symbols, state the hypotheses for this test.
b. Locate the critical region for $\alpha = .05$.
c. Compute the t statistic for these data.
d. What decision should the researcher make.

4. An educator would like to assess the effectiveness of a new instructional program for reading. The control group consists of $n = 4$ second-grade students who are provided with the traditional instruction used at the school. Another sample of $n = 4$ second-graders receives the experimental instruction. The subjects in the experimental group are matched one-to-one with the subjects in the control group based on their reading achievement test scores from the previous year. After six months, both groups are tested with a standard reading exam. The data are as follows:

Matched Pair	Control	Experimental
A	10	12
B	15	25
C	13	15
D	18	20

On the basis of these data can the educator conclude that the special program has a significant effect on reading scores? Test at the .05 level.

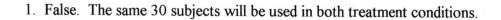

ANSWERS TO SELF-TEST

True/False Answers

1. False. The same 30 subjects will be used in both treatment conditions.

2. True

3. False. A sample of n = 19 produces df = 18.

4. True

5. True

6. True

7. True.

8. False. The larger the mean difference, the larger the value for t.

9. True

10. True

Multiple-Choice Answers

1. b 2. a 3. d 4. a 5. b 6. b 7. b 8. c 9. a 10. d

Other Answers

1. a. An independent-measures design would require two separate samples of n = 30 for a total of 60 subjects.
   b. A repeated-measures design would require only one sample of n = 30 subjects.
   c. A matched-subjects design would require two matched samples of n = 30 for a total of 60 subjects.

2.

Subject	1st	2nd	D	
#1	8	10	+2	
#2	6	12	+6	$\Sigma D = 20$
#3	10	7	-3	
#4	9	17	+8	$\bar{D} = 20/5 = 4$
#5	7	14	+7	

3. a. H_0: $\mu_D = 0$ (no effect). H_1: $\mu_D \neq 0$
   b. With df = 15 and $\alpha = .05$, the critical value are $t = \pm 2.131$
   c. The sample variance is $s^2 = 400$, the standard error is 5 points, and $t(15) = 4.2$.
   d. Reject the null hypothesis. The diet drug has a significant effect on activity.

4. The null hypothesis states that there is no difference between the traditional program and the new program. With df = 3, the critical t value is 3.182. For these data, $\overline{D} = 4$, SS = 48, and t = 2. Fail to reject H_0.

Chapter 12

Estimation

Chapter 12 introduces the inferential process of estimation. In general terms, estimation uses a sample statistic as the basis for estimating the value of the corresponding population parameter. Although estimation and hypothesis testing are similar in many respects, they are complementary inferential processes. A hypothesis test is used to determine whether or not a treatment has an effect, while estimation is used to determine how much effect.

This complementary nature occurs because estimation is often used after a hypothesis test that resulted in rejecting the null hypothesis. In this situation, the hypothesis test has established that a treatment effect exists and the next logical step is to determine how much effect. It is also common to use estimation in situations where a researcher simply wants to learn about an unknown population. In this case, a sample is selected from the population and the sample data are then used to estimate the population parameters.

You should keep in mind that even though estimation and hypothesis testing are inferential procedures, these two techniques differ in terms of the type of question they address. A hypothesis test, for example, addresses the somewhat academic question concerning the <u>existence</u> of a treatment effect. Estimation, on the other hand, is directed toward the more practical question of <u>how much</u> effect. When a hypothesis test established that a treatment has a "statistically significant" effect, it simply means that the observed treatment effect is greater than would be expected by chance. You should note that this conclusion does not necessarily imply anything about the actual size of the effect. A new medication, for example, may consistently and reliably reduce ' blood pressure by 1%. Although this effect may be statistically significant, it may have little or no practical significance for treating high blood pressure. By using estimation, researchers are able to determine the magnitude of the treatment effect. Often the size of a treatment effect is more important than the mere existence of an effect, especially when the treatment is being considered for practical application.

There is an estimation procedure that accompanies each of the four hypothesis tests presented in the preceding four chapters. The estimation process begins with the same z or t statistic that is used for the corresponding hypothesis test. These all have the same conceptual structure:

$$z \text{ or } t = \frac{\text{sample statistic - unknown parameter}}{\text{standard error}}$$

The basic estimation formula is obtained by solving this equation for the unknown parameter:

$$\text{unknown parameter} = \text{statistic} \pm (z \text{ or } t)(\text{standard error})$$

To use this equation, you first must obtain a value for z or t by estimating the location of the data within the appropriate distribution. The estimated value for z or t is then substituted in the equation along with the known values for the statistic and the standard error. Then the equation is solved for the unknown parameter. For a point estimate, a value of zero is used for z or t, yielding a single value for your estimate of the unknown population parameter. For interval estimates, you first select a level of confidence and then find the appropriate interval range of z or t values in the unit normal table and t distribution table, respectively.

1. You should be able to use sample data to make a point estimate or an interval estimate of an unknown population mean using either a z-score (when σ is known) or a t statistic (when σ is unknown).

2. You should be able to make a point estimate or an interval estimate of a population mean difference using sample data from an independent-measures experiment or from a related-samples experiment.

3. You should understand how the size of a sample influences the width of a confidence interval.

4. You should understand how the level of confidence (the % confidence) influences the width of a confidence interval.

NEW TERMS AND CONCEPTS

The following terms were introduced in this chapter. You should be able to define or describe each term and, where appropriate, describe how each term is related to other terms in the list.

estimation   The inferential process of using a sample statistic to estimate the value of an unknown population parameter.

point estimate

A single number is used to estimate the value of an unknown parameter. The result is a precise estimate, but without confidence.

interval estimate

A range of values is used to estimate an unknown parameter. The result is a less precise estimate but there is a gain in confidence.

confidence interval

An interval estimate that is described in terms of the level (percentage) of confidence in the accuracy of the estimation.

NEW FORMULAS

$$\mu = \overline{X} \pm z\sigma_{\overline{X}}$$

$$\mu = \overline{X} \pm ts_{\overline{X}}$$

$$\mu_D = \overline{D} \pm ts_{\overline{D}}$$

$$(\mu_1 - \mu_2) = (\overline{X}_1 - \overline{X}_2) \pm ts_{\overline{X}_1 - \overline{X}_2}$$

The following examples will be used to demonstrate the step-by-step procedures for estimation with z-scores and with the t statistic.

A researcher begins with a normal population with $\mu = 60$ and $\sigma = 8$. The researcher is evaluating a specific treatment that is expected to increase scores. The treatment is administered to a sample of $n = 16$ individuals, and the mean for the treated sample is $\overline{X} = 66$.

<u>Estimation with z-scores</u>: For our example, the researcher does not know the mean for the population after treatment. The researcher expects that the treated population will have a mean greater than 60 because the treatment is expected to increase scores. However, the only available information comes from the sample mean which can be used to estimate the mean for the unknown population. Because the population standard deviation is known, the z-score formula for estimation is appropriate.

Step 1: Begin with the basic formula for estimation. Remember, this is simply the regular z-score formula that has been solved for μ.

$$\mu = \overline{X} \pm z\sigma_{\overline{X}}$$

Step 2: Determine whether you are computing a point estimate or an interval estimate. If you want an interval, you must specify a level of confidence. We will compute a point estimate and a 90% confidence interval for the unknown population mean.

Step 3: Find the appropriate z-score values to substitute in the equation. For a point estimate, always use $z = 0$ which is the point in the exact middle of the distribution. For a 90% confidence interval, we want the z-score values that form the boundaries for the middle 90% of the distribution. These values are obtained from the unit normal table. You should find that 90% of a normal distribution is contained between $z = +1.65$ and $z = -1.65$.

Step 4: Compute $\overline{X}$ and $\sigma_{\overline{X}}$ from the sample data. For this example we are given $\overline{X} = 66$ and you can compute

$$\sigma_{\overline{X}} = \sigma/\sqrt{n} = 8/\sqrt{16} = 8/4 = 2$$

Step 5: Substitute the appropriate values in the estimation equation.

point estimate	interval estimate
$\mu = 66 \pm 0$	$\mu = 66 \pm (1.65)(2)$
$\mu = 66$	$\mu = 66 \pm 3.30$

Estimate μ between
62.70 and 69.30

Estimation with the t statistic: For our example, the population standard deviation was given. But if the value for σ is unknown, it is still possible to estimate the mean for the population after treatment. Suppose that our researcher simply knows that the original population mean is $\mu = 60$ and that the treatment is expected to produce an increase in the scores. The treatment is administered to a sample of $n = 16$ individuals and produces a sample mean of $\overline{X} = 66$ with $SS = 1215$. Because the population standard deviation is not known, the t statistic formula for estimation is appropriate.

Step 1: Begin with the basic formula for estimation. Again, this is simply the regular t formula that has been solved for μ.

$$\mu = \overline{X} \pm ts_{\overline{X}}$$

Step 2: Determine whether you are computing a point estimate or an interval estimate. If you want an interval, you must specify a level of confidence. We will compute a point estimate and a 90% confidence interval for the unknown population mean.

Step 3: Find the appropriate t values to substitute in the equation. For a point estimate, always use t = 0 which is the point in the exact middle of the distribution. For a 90% confidence interval, we want the t values that form the boundaries for the middle 90% of the distribution. These values are obtained from the t distribution table using df = n - 1 = 15. You should find that with df = 15, 90% of the t distribution is contained between t = +1.753 and t = -1.753.

Step 4: Compute $\overline{X}$ and $s_{\overline{X}}$ from the sample data. For this example we are given $\overline{X}$ = 66 and you can compute

$$s^2 = \frac{SS}{n-1} = \frac{1215}{15} = 81$$

$$s_{\overline{X}} = \sqrt{\frac{s^2}{n}} = \sqrt{\frac{81}{16}} = \frac{9}{4} = 2.25$$

Step 5: Substitute the appropriate values in the estimation equation.

point estimate	interval estimate
$\mu = 66 \pm 0$	$\mu = 66 \pm (1.753)(2.25)$
$\mu = 66$	$\mu = 66 \pm 3.94$

Estimate μ between
62.06 and 69.94

HINTS AND CAUTIONS

1. When computing a confidence interval after a hypothesis test, many students incorrectly take the z-score or t value that was computed in the hypothesis test and use this value in the estimation equation. Remember, the z-score or t value in the estimation equation is determined by the level of confidence. For example, an 80% confidence interval with z-scores always uses $z = \pm 1.28$ no matter what z-score was obtained from the hypothesis test.

2. When trying to locate the appropriate t value for a confidence interval, remember to use the "proportion in two tails" column of the t table. For an 80% confidence interval, for example, there would be 20% of the distribution left in the two tails and you should use the .20 column for two tails of the distribution.

SELF-TEST

True/False Questions

1. To gain precision in an estimate, you must lose some confidence.

2. In general terms, the purpose for estimation is to determine whether or not a treatment has an effect.

3. The sample mean will always be exactly in the center of a confidence interval that is estimating the value of the population mean.

4. If all other factors are held constant, then a 95% confidence interval will be wider than an 80% confidence interval.

5. If all other factors are held constant, then a confidence interval based on a sample of n = 10 will be wider than a confidence interval based on a sample of n = 20.

6. For a point estimate, you always use a value of t = 0 (or z = 0) in the estimation equation.

7. Usually it would not be reasonable to use estimation after a hypothesis test where the decision was to reject the null hypothesis.

8. For a confidence interval based on a t statistic, the larger the sample variance, the wider the confidence interval. (Assume that all other factors are held constant.)

9. If all other factors are held constant, the width of the confidence interval will decrease as the value of df increases.

10. A researcher is using a sample of n = 16 scores to construct a 90% confidence interval for the population mean. The correct t statistic values for this interval are t = ±2.131,

Multiple-Choice Questions

1. If all other factors are held constant, which confidence level will produce the smallest width for a confidence interval?
    a. 99%
    b. 90%
    c. 75%
    d. 60%

2. A sample has a mean of $\overline{X} = 53$. If this sample is used to make a point estimate of the population mean, then the point estimate would be

    a. Cannot answer without knowing the sample size

    b. Cannot answer without knowing the population standard deviation or the sample variance.

    c. All of the above

    d. $\mu = 53$

3. A researcher uses a sample of $n = 25$ scores to construct a 90% confidence interval to estimate an unknown population mean. The correct interpretation of the confidence interval is

    a. If another sample of $n = 25$ scores was selected, there is a 90% probability that the sample mean will be in the interval

    b. There is a 90% probability that the unknown population mean is in the interval.

    c. All of the above

    d. None of the above

4. Assuming that all other factors are held constant, which of the following would produce the widest confidence interval.

    a. An 80% confidence interval based on a sample of $n = 4$ scores

    b. An 80% confidence interval based on a sample of $n = 25$ scores

    c. A 90% confidence interval based on a sample of $n = 4$ scores

    d. A 90% confidence interval based on a sample of $n = 25$ scores

5. What are the correct values of t to use for a 90% confidence interval based on a sample of $n = 9$ scores?

    a. $t = \pm 1.397$

    b. $t = \pm 1.383$

    c. $t = \pm 1.860$

    d. $t = \pm 1.833$

6. A sample of n = 4 scores is selected from a population with an unknown mean. The sample has a mean of $\overline{X} = 40$ and a variance of $s^2 = 16$. Which of the following is the correct 90% confidence interval for μ?

    a. $\mu = 40 \pm 2.353(4)$

    b. $\mu = 40 \pm 1.638(4)$

    c. $\mu = 40 \pm 2.353(2)$

    d. $\mu = 40 \pm 1.638(2)$

7. A researcher knows that 4-year-old girls tend to have better verbal skills than 4-year-old boys. To determine how much better, the researcher obtains a sample of boys and a sample of girls and gives each child a verbal ability test. The scores from the test are used to estimate the mean difference. Which estimation equation should the researcher use?

    a. The z-score equation

    b. The single-sample t equation

    c. The independent-measures t equation

    d. The repeated-measures t equation

8. Although it is known that a particular therapy is effective for reducing depression, a researcher would like to determine how much effect the therapy has. A sample of depressed patients is obtained and each individual is given a depression test. After four weeks of therapy, their depression is measured again. The scores from the depression tests are used to estimate the mean difference in depression. Which estimation equation should the researcher use?

    a. The z-score equation

    b. The single-sample t equation

    c. The independent-measures t equation

    d. The repeated-measures t equation

9. A researcher would like to determine how much difference there is between two treatment conditions. A sample of n = 10 people is tested in the first treatment and produces a mean score of $\overline{X}$ = 42 with SS = 160. A second sample of n = 10 people is tested in the second treatment and produces $\overline{X}$ = 48 with SS = 200. If the researcher wants to construct a 90% confidence interval for the mean difference, then what t values should be used in the estimation equation?

    a. t = ± 1.833
    b. t = ± 1.729
    c. t = ± 1.734
    d. t = ± 2.101

10. A researcher would like to determine how much difference there is between two treatment conditions. A sample of n = 10 people is tested in the first treatment and produces a mean score of $\overline{X}$ = 48 with SS = 200. A second sample of n = 10 people is tested in the second treatment and produces $\overline{X}$ = 42 with SS = 160. If the researcher used these data to make a point estimate of the population mean difference, then what would the estimate be?

    a. $\mu_1 - \mu_2 = 6$
    b. $\mu_1 - \mu_2 = 40$
    c. $\mu_1 - \mu_2 = 42$
    d. $\mu_1 - \mu_2 = 48$

Other Questions

1. Estimation often is used after a hypothesis test. Although these two inferential techniques involve many of the same calculations, they are intended to answer different questions.

    a. In general terms, what information is provided by a hypothesis test and what information is provided by estimation?
    b. Explain why it would not be appropriate do use estimation after a hypothesis test where the decision was "fail to reject" the null hypothesis.

2. A sample of n = 16 scores is obtained from a normal population with σ = 12. The sample mean is $\overline{X}$ = 43.

    a. Use these data to make a point estimate of the population mean.

    b. Make an interval estimate of μ so that you are 90% confident that the mean is in your interval.

3. A researcher studying animal learning is investigating the effectiveness of a new training procedure. Under regular circumstances, rats require an average of μ = 35 errors before they master a standard problem-solving task. The researcher tests the effectiveness of the new procedure using a sample of n = 25 rats, and obtains a sample mean of $\overline{X}$ = 29 errors with SS = 2400.

    a. Make a point estimate of the population mean number of errors using the new training procedure.

    b. Make an interval estimate of the population mean so that you are 90% confident that the new mean is in your interval.

4. A high school counselor has designed a three-week course to provide students with special training in study skills and note taking. To evaluate the effectiveness of this course, a sample of n = 36 sophomores is obtained. The counselor records the fall term grade average for each student. These students then take the special course before the spring term begins, and the counselor records each student's average for the spring term. On average, these students scored $\overline{D}$ = 8 points higher during the spring term with SS = 1260.

    a. On the basis of these data, can the counselor conclude that the special course has a significant effect on student performance? Test at the .05 level of significance.

    b. Use the data to make a point estimate of how much improvement results from taking the special course.

    c. Make an interval estimate of the mean improve ment so that you are 90% confident that the true mean improvement is contained in your interval.

5. The government has developed a pamphlet listing driving tips that are designed to promote fuel economy. To test the effectiveness of these tips, a sample of twenty families is obtained, all with identical cars and similar driving habits. Ten of these families are given the pamphlet and instructions to follow the driving tips carefully. The other ten families are simply instructed to monitor their gas mileage for a two-week period. At the end of two weeks, the gas mileage figures for both groups are as follows:

Experimental Group with Pamphlet		Control Group no Pamphlet	
25.1	28.7	24.0	23.6
22.8	25.0	25.3	25.4
27.2	29.1	24.9	22.3
28.4	27.5	21.6	23.2
26.9	30.1	24.1	26.2

a. Do these data indicate that the driving tips have a significant effect on gas mileage? Test at the .05 level of significance.
b. Use the data to make a point estimate of how much improvement in gas mileage results from following the tips.
c. Use the data to construct a 90% confidence interval to estimate how much effect the driving tips have on gas mileage.

====

ANSWERS TO SELF-TEST

====

True/False Answers

1. True

2. False. A hypothesis test is used to determine whether or not a treatment has an effect. Estimation is used to determine how much effect.

3. True

4. True

5. True

6. True

7. False. Rejecting the null hypothesis means that you have concluded that the treatment does have an effect. Estimation would be the next logical step to determine how much effect.

8. True

9. True

10. False. The correct t values are $t = \pm 1.753$

Multiple-Choice Answers

1. d   2. d   3. b   4. c   5. c   6. c   7. c   8. d   9. c   10. a

Other Answers

1. a. A hypothesis test is intended to determine whether or not a treatment effect exists. Estimation is used to determine how much effect.

   b. If the decision from the hypothesis test is "fail to reject H_0 "then the data do not provide sufficient evidence to conclude that there is any treatment effect. In this case, it would not be reasonable to use estimation in an attempt to determine "how much" effect exists.

2. a. Use $\overline{X} = 43$ as the point estimate of μ.
   b. The 90% confidence interval extends from 38.05 to 47.95.

3. a. Use the sample mean, $\overline{X} = 29$ as the point estimate of μ.
   b. The 90% confidence interval would be $\mu = 29 \pm (1.711)(2)$
      The interval extends from 32.422 to 25.578.

4. a. With $s^2 = 36$ and $s_{\overline{X}} = 1$, the data have a t statistic of $t = 8.00$. Reject H_0 and conclude that the course has a significant effect.
   b. Use the sample mean, $\overline{D} = 8$ points, as the point estimate of μ_D.
   c. The 90% confidence interval would be $\mu_D = 8 \pm (1.684)(1)$
      which gives an interval from 6.316 to 9.684.

5. a. For the experimental group, $\overline{X} = 27.0$ with $s = 2.40$. For the control group, $\overline{X} = 24.06$ with $s = 1.44$. The t statistic is $t = 3.32$. Reject H_0.
   b. The sample mean difference, 2.94 miles per gallon, is the best point estimate.
   c. The 90% confidence interval would be $\mu_1 - \mu_2 = 2.94 \pm (1.734)(0.88)$
      which gives an interval estimate between 4.47 and 1.41.

Chapter 13

Introduction to
Analysis of Variance

CHAPTER SUMMARY

Chapter 13 presents the general logic and basic formulas for the hypothesis testing procedure known as analysis of variance (ANOVA). The purpose of ANOVA is much the same as the t tests presented in the preceding three chapters: the goal is to determine whether the mean differences that are obtained for sample data are sufficiently large to justify a conclusion that there are mean differences between the populations from which the samples were obtained. The difference between ANOVA and the t tests is that ANOVA can be used in situations where there are <u>two or more</u> means being compared, whereas the t tests are limited to situations where only two means are involved.

Analysis of variance is necessary to protect researchers from excessive risk of a Type I error in situations where a study is comparing more than two population means.

These situations would require a series of several t tests to evaluate all of the mean differences. (Remember, a t test can compare only 2 means at a time.) Although each t test can be done with a specific α-level (risk of Type I error), the α-levels accumulate over a series of tests so that the final, "experimentwise," α-level can be quite large. ANOVA allows researcher to evaluate all of the mean differences in a single hypothesis test using a single α-level and, thereby, keeps the risk of a Type I error under control no matter how many different means are being compared.

Although ANOVA can be used in a variety of different research situations, this chapter presents only independent-measures designs involving only one independent variable.

Hypothesis Tests with Analysis of Variance (ANOVA): The test statistic for ANOVA is an F-ratio, which is a ratio of two sample variances. In the context of ANOVA, the sample variances are called mean squares, or MS values. The top of the F-ratio $MS_{between}$ measures the size of mean differences between samples. The bottom of the ratio MS_{within} measures the magnitude of differences that would be expected by chance or sampling error.

Thus, the F-ratio has the same basic structure as the independent-measures t statistic presented in Chapter 10.

$$F = \frac{\text{obtained mean differences}}{\text{differences expected by chance (error)}} = \frac{MS_{between}}{MS_{within}}$$

A large value for the F-ratio indicates that the obtained sample mean differences are greater than would be expected by chance.

Each of the sample variances, MS values, in the F-ratio is computed using the basic formula for sample variance:

$$\text{sample variance} = MS = \frac{SS}{df}$$

To obtain the SS and df values, you must go through an analysis that separates the total variability for the entire set of data into two basic components: between-treatment variability (which will become the numerator of the F-ratio), and within-treatment variability (which will be the denominator). The two components of the F-ratio can be described as follows:

Between-Treatments Variability: $MS_{between}$ measures the size of the differences between the sample means. For example, suppose that three treatments, each with a sample of $n = 5$ subjects, have means of $\overline{X}_1 = 1$, $\overline{X}_2 = 2$, and $\overline{X}_3 = 3$. Notice that the three means are different; that is, they are variable. By computing the variance for the three means we can measure the size of the differences. Although it is possible to compute a variance for the set of sample means, it usually is easier to use the total, T, for each sample instead of the mean, and compute variance for the set of T values. Logically, the differences (or variance) between means can be caused by two sources:

    1. Treatment Effects: If the treatments have different effects, this could cause the scores in one treatment to be higher (or lower) than the scores in another treatment.

    2. Chance: If there is no treatment effect at all, you would still expect some differences between samples. For example, the individual subjects in one treatment are different from the subjects in another treatment. These individual differences could cause the scores to be different from one treatment to another.

Within-Treatments Variability: MS_{within} measures the size of the differences between scores within each of the samples. MS_{within} is calculated by finding the variance, s^2 for each sample, and then pooling the sample variances exactly as we did to find pooled variance for the independent-measures t statistic.

$$MS_{within} = \frac{SS_{within}}{df_{within}} = \frac{\Sigma SS}{\Sigma df} = \frac{SS_1 + SS_2 + SS_3 + ...}{df_1 + df_2 + df_3 + ...}$$

which is the same structure as

$$\text{pooled variance} = \frac{SS_1 + SS_2}{df_1 + df_2}$$

Because all the individuals in a sample receive exactly the same treatment, any differences (or variance) within a sample can be caused by only one source:

    1. Chance: The unpredictable differences that exist between individual scores are simply consider to be chance differences.

Considering these sources of variability, the structure of the F-ratio becomes,

$$F = \frac{\text{treatment effect} + \text{chance}}{\text{chance}}$$

When the null hypothesis is true and there are no differences between treatments, the F-ratio is balanced. That is, when the "treatment effect" is zero, the top and bottom of the F-ratio are measuring the same variance. In this case, you should expect an F-ratio near 1.00. When the sample data produce an F-ratio near 1.00, we will conclude that there is no significant treatment effect.

On the other hand, a large treatment effect will produce a large value for the F-ratio. Thus, when the sample data produce a large F-ratio we will reject the null hypothesis and conclude that there are significant differences between treatments.

To determine whether an F-ratio is large enough to be significant, you must select an α-level, find the df values for the numerator and denominator of the F-ratio, and consult the F-distribution table to find the critical value.

ANOVA and Post Tests: The null hypothesis for ANOVA states that for the general population there are no mean differences among the treatments being compared;
$H_0: \mu_1 = \mu_2 = \mu_3 = ...$

When the null hypothesis is rejected, the conclusion is that there are significant mean differences. However, the ANOVA simply establishes that differences exist, it does not indicate exactly which treatments are different. With more than two treatments, this creates a problem. Specifically, you must follow the ANOVA with additional tests, called post tests, to determine exactly which treatments are different and which are not. The Scheffe test is an example of a post tests. These tests are done after an ANOVA where H_O is rejected with more than two treatment conditions. The tests compare the treatments, two at a time, to test for mean differences.

ANOVA and the Independent-Measures t Test: When a research study compares only two treatments using an independent-measures design, the data may be analyzed using either an analysis of variance or an independent-measures t test. The two procedures evaluate exactly the same hypotheses and always will reach exactly the same conclusion. In addition, the t statistic obtained from the data and the F-ratio obtained from the data are directly related by $F = t^2$. The critical values for t and F are also related by $F = t^2$. Finally, the df value for the t statistic (associated with the pooled variance) will be equal to the df value for MS_{within} (the denominator of the F-ratio which also is measuring pooled variance).

LEARNING OBJECTIVES

1. You should be familiar with the purpose, terminology, and special notation of analysis of variance.

2. You should be able to perform an analysis of variance for the data from a single-factor, independent-measures experiment.

3. You should recognize when post hoc tests are necessary and you should be able to complete an analysis of variance using the Scheffe' post hoc test.

4. You should be able to report the results of an analysis of variance using either a summary table or an F-ratio (including df values). Also, you should be able to understand and interpret these reports when they appear in scientific literature.

NEW TERMS AND CONCEPTS

The following terms were introduced in this chapter. You should be able to define or describe each term and, where appropriate, describe how each term is related to other terms in the list.

factor	In analysis of variance, an independent variable (or quasi-independent variable) is called a factor.
levels of a factor	The specific conditions or values that are used to represent the factor are called levels.
F-ratio	The test statistic for analysis of variance is called an F-ratio and compares the differences (variance) between treatments with the differences (variance) that are expected by chance.
MS (Mean Square)	In analysis of variance, a sample variance is called a mean square or MS, indicating that variance measures the mean of the squared deviations.

post hoc test	A test that is conducted after an ANOVA with more than two treatment conditions where the null hypothesis was rejected. The purpose of post hoc tests is to determine exactly which treatment conditions are significantly different.
between treatments SS, df, MS	Values used to measure and describe the differences between treatments (mean differences).
within treatments SS, df, MS	Values used to measure and describe the differences inside the treatment conditions. These differences are assumed to measure chance or error variability.
total SS and df	Values used to measure and describe the total amount of variability for the entire set of data.

NEW FORMULAS

$$SS_{total} = \Sigma X^2 - \frac{G^2}{N} \qquad df_{total} = N - 1$$

$$SS_{between} = \Sigma \frac{T^2}{n} - \frac{G^2}{N} \qquad df_{between} = k - 1$$

$$SS_{within} = \Sigma SS_{each\ treatment} \qquad df_{within} = N - k$$

$$MS = \frac{SS}{df} \qquad\qquad F = \frac{MS_{between}}{MS_{within}}$$

STEP BY STEP

Analysis of Variance: Analysis of variance is a hypothesis testing technique that is used to determine whether there are differences among the means of two or more populations. In this chapter we examined ANOVA for an independent-measures experiment, which means that the data consist of a separate sample for each treatment condition (or each population). Before you begin the actual analysis, you should complete all the preliminary calculations with the data, including T and SS for each sample and G and ΣX^2 for the entire set of scores. The following example will be used to demonstrate ANOVA.

A researcher has obtained three different samples representing three populations. The data are presented below.

Sample 1	Sample 2	Sample 3	
0	6	6	
4	8	5	G = 60
0	5	9	
1	4	4	$\Sigma X^2 = 356$
0	2	6	
T = 5	T = 25	T = 30	
SS = 12	SS = 20	SS = 14	

Step 1: State the hypotheses and select an alpha level. The null hypothesis states that there are no mean differences among the three populations.

$H_0: \mu_1 = \mu_2 = \mu_3$

Remember, we generally do not try to list specific alternatives, but rather state a generic alternative hypothesis.

H_1: At least one population mean is different from the others

For this test we will use $\alpha = .05$.

Step 2: Locate the critical region. With $k = 3$ samples, the numerator of the F-ratio will have $df_{between} = k - 1 = 2$. There are $n = 5$ scores in each sample. Within each sample there are $n - 1 = 4$ degrees of freedom, and summing across all three samples gives $df_{within} = 4 + 4 + 4 = 12$ for the denominator of the F-ratio. Thus, the F-ratio for this analysis will have $df = 2,12$.

Sketch the entire distribution of F-ratios with $df = 2,12$ and locate the extreme 5% of the distribution. The critical F value is 3.88.

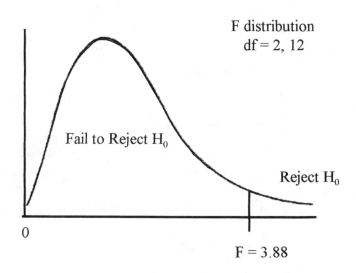

Step 3: Compute the test statistic. It is best to work through the calculations in a systematic way. Compute all three parts of the analysis (total, between, and within) and check that the two components add to the total.

$$SS_{total} = \Sigma X^2 - \frac{G^2}{N} = 356 - \frac{(60)^2}{15}$$

$$= 356 - 240$$

$$= 116$$

$$SS_{bet} = \Sigma \frac{T^2}{n} - \frac{G^2}{N} = \frac{(5)^2}{5} + \frac{(25)^2}{5} + \frac{(30)^2}{5} - \frac{(60)^2}{15}$$

$$= 5 + 125 + 180 - 240$$

$$= 70$$

$$SS_{within} = \Sigma SS = 12 + 20 + 14 = 46$$

(Check that $116 = 70 + 46$)

We have already found $df_{between}$ and df_{within}. To complete the analysis of df, compute

$$df_{total} = N - 1 = 15 - 1 = 14$$

(Check that $14 = 12 + 2$)

Next, compute the two variances (Mean Squares) that will form the F-ratio.

$$MS_{between} = \frac{SS_{between}}{df_{between}} = \frac{70}{2} = 35$$

$$MS_{within} = \frac{SS_{within}}{df_{within}} = \frac{46}{12} = 3.83$$

Finally, the F-ratio for these data is,

$$F = \frac{MS_{between}}{MS_{within}} = \frac{35}{3.83} = 9.14$$

Step 4: Make decision. The F-ratio for these data is in the critical region. The numerator is more than 9 times larger than the denominator which indicates a significant treatment effect. Reject H_0 and conclude that there are differences among the means of the three populations.

HINTS AND CAUTIONS

1. It may help you to understand analysis of variance if you remember that measuring variance is conceptually the same as measuring differences. The goal of the analysis is to determine whether the mean differences in the data are larger than would be expected by chance.

2. The formulas for SS between treatments and SS within treatments, and their role in the F-ratio may be easier to remember if you look at the similarities between the independent-measures t formula and the F-ratio formula.

    a. The numerator of the t statistic measures the difference between the two sample means, $(\overline{X}_1 - \overline{X}_2)$. The numerator of the F-ratio also looks at differences between treatments by computing the variability for the treatment totals.

    b. The standard error in the denominator of the t statistic is computed by first pooling the two sample variances. This calculation uses the SS values from each of the two separate samples. The denominator of the F-ratio also uses the SS values from each of the separate samples to compute SS within treatments. In fact, when there are only two treatment conditions, the pooled variance from the t statistic is equivalent to the MS within treatments from the F-ratio.

SELF-TEST

True/False Questions

1. It is impossible to obtain a negative value for an F- ratio.

2. If the null hypothesis is true, the F-ratio for ANOVA is expected (on average) to have a value of 1.00.

3. An F-ratio of zero $(F = 0)$ indicates that all of the separate samples have exactly the same mean.

4. In an analysis of variance, if all other factors are held constant, the larger the differences between the sample means, the bigger the value for the F-ratio.

5. The F-ratio for an analysis of variance comparing two treatments with n = 10 in each treatment would have df = 2, 18.

6. The critical region for an analysis of variance is located entirely in one tail of the distribution.

7. An analysis of variance comparing 3 treatment conditions produces $MS_{between} = 10$. For this analysis, $SS_{between} = 30$.

8. Post tests are used after an analysis of variance to determine whether or not a Type I error was made during the ANOVA.

9. Post tests are done after an analysis of variance when the statistical decision is "Fail to Reject H_o."

10. Post tests are <u>not</u> necessary if there are only two treatments in the original ANOVA.

Multiple-Choice Questions

1. A researcher uses an analysis of variance to test for mean differences among three treatment conditions using a sample of n = 10 subjects in each treatment. The F-ratio from this analysis would have
     a. df = 29
     b. df = 2, 29
     c. df = 3, 27
     d. df = 2, 27

2. A researcher reports an F-ratio with df = 3, 36 from an independent-measures research study. Based on the df values, how many treatments were compared in the study and what was the total number of subjects participating in the study?

    a. 3 treatments and 37 subjects
    b. 4 treatments and 36 subjects
    c. 4 treatments and 40 subjects
    d. 2 treatments and 35 subjects

3. A researcher obtains an F-ratio of F = 4.00 from an independent-measures research study comparing 2 treatment conditions. If the researcher had used an independent-measures t statistic to evaluate the data, what value would be obtained for the t statistic?

    a. t = 2
    b. t = 4
    c. t = 16
    d. Cannot determine from the information given

4. In general, the largest F-ratio will be obtained when the differences between sample means are _____ and the magnitudes of the sample variances are _____.

    a. small, small
    b. small, large
    c. large, small
    d. large, large

5. An independent-measures research study compares three treatment conditions using a sample of n = 5 in each treatment. For this study, the three sample totals are, $T_1 = 5$, $T_2 = 10$, and $T_3 = 15$. What value would be obtained for $SS_{between}$?

    a. 1
    b. 5
    c. 10
    d. 15

6. An independent-measures research study compares three treatment conditions using a sample of $n = 10$ in each treatment. For this study, the three samples have $SS_1 = 10$, $SS_2 = 20$, and $SS_3 = 15$. What value would be obtained for MS_{within}?

    a. 45/30

    b. 45/29

    c. 45/27

    d. 45/3

7. Which of the following is not normally calculated as part of an analysis of variance?

    a. SS_{total}

    b. SS_{within}

    c. MS_{total}

    d. MS_{within}

8. An independent-measures research study compares three treatment conditions using a sample of $n = 5$ in each treatment. For this study, the three sample totals are, $T_1 = 5$, $T_2 = 10$, and $T_3 = 15$, and $\Sigma X^2 = 75$ for the entire set of scores.. What value would be obtained for SS_{total}?

    a. 1

    b. 5

    c. 10

    d. 15

9. The purpose for post tests is

    a. To determine whether or not a Type I error was committed.

    b. To determine how much difference exists between the treatments

    c. To determine which treatments are significantly different

    d. None of the above

10. Under what circumstances are post test necessary?
   a. When you reject the null hypothesis with exactly 2 treatment conditions.
   b. When you reject the null hypothesis with more than 2 treatment conditions
   c. When you fail to reject the null hypothesis with exactly 2 treatment conditions.
   d. When you fail to reject the null hypothesis with more than 2 treatment conditions.

Other Questions

1. A researcher conducts an experiment comparing four treatment conditions with a separate sample of $n = 5$ in each treatment. An analysis of variance is used to evaluate the data, and the results of the ANOVA are presented in the table below. Complete all missing values in the table.

Source	SS	df	MS	
Between Treatments	12	__	__	$F = 2.00$
Within Treatments	__	__	__	
Total	__	__		

2. The following data summarize the results of two experiments. Each experiment compares three treatment conditions, and each experiment uses separate samples of $n = 10$ for each treatment.

Experiment A			Experiment B		
Treatment			Treatment		
1	2	3	1	2	3
$\overline{X} = 1$	$\overline{X} = 3$	$\overline{X} = 5$	$\overline{X} = 1$	$\overline{X} = 10$	$\overline{X} = 20$
$s = 15$	$s = 12$	$s = 18$	$s = 3$	$s = 5$	$s = 4$

Just looking at the data - without doing any calculations - answer each of the following questions.

    a. Which experiment will produce the larger $MS_{between}$?

    b. Which experiment will produce the larger MS_{within}?

    c. Which experiment will produce the larger F-ratio?

3.   Use an analysis of variance with $\alpha = .05$ to determine whether the following data provide evidence of any significant differences among the three treatments.

Treatments

I	II	III	
0	4	1	$G = 30$
2	6	0	
2	1	3	$\Sigma X^2 = 114$
0	5	1	
1	4	0	
T = 5	T = 20	T = 5	
SS = 4	SS = 14	SS = 6	

4. The following data are from two separate samples.

    a. Use an analysis of variance with $\alpha = .05$ to determine whether these data provide evidence for a significant difference between the two population means.

    b. If you had used a t test instead of ANOVA, what value would you have obtained for the t statistic?

Sample 1	Sample 2
n = 10	n = 10
$\overline{X} = 3$	$\overline{X} = 5$
SS = 200	SS = 160

ANSWERS TO SELF-TEST

True/False Answers

1. True

2. True

3. True

4. True

5. False. The F-ratio would have df = 1, 18.

6. True

7. False. $SS_{between} = 20$.

8. False. Post tests are done to determine exactly which treatment are significantly different.

9. False. Post tests are done when the null hypothesis is rejected.

10. True

Multiple-Choice Answers

1. d   2. c   3. a   4. c   5. c   6. c   7. c   8. d   9. c   10. b

Other Answers

1.

Source	SS	df	MS	
Between Treatments	12	3	4	$F = 2.00$
Within Treatments	32	16	2	
Total	44	19		

2. a. Experiment B has larger mean differences and will produce a larger $MS_{between}$.

   b. Experiment A has larger sample standard deviations (more variance within samples) and will produce a larger MS_{within}.

   c. Experiment B has larger differences between treatments and smaller variability within treatments. This combination will produce a larger F-ratio.

3. The ANOVA is summarized as follows:

Source	SS	df	MS	
Between Treatments	30	2	15	$F = 7.50$
Within Treatments	24	12	2	
Total	54	14		

With df = 2, 12 the critical value is F = 3.88. Reject H_0 and conclude that there are significant differences among the three treatments.

4. a. With df = 1,18 the critical value for F is 4.41. For these data,

Source	SS	df	MS	
Between Treatments	20	1	20	$F = 1.00$
Within Treatments	360	18	20	
Total	380	19		

   Fail to reject H_0.

   b. The t test would produce a t statistic of $t = \sqrt{F} = 1.00$.

Chapter 14

Two-Factor
Analysis of Variance

CHAPTER SUMMARY

Chapter 14 extends analysis of variance to research designs that involve two independent variables. In the context of ANOVA, an independent variable (or a quasi-independent variable) is called a factor, and research studies with two factors are called <u>factorial designs</u> or simply <u>two-factor designs</u>. The two factors are identified as A and B, and the structure of a two-factor design can be represented as a matrix with the levels of factor A determining the rows and the levels of factor B determining the columns. For example, a researcher studying the effects of heat and humidity on performance could use the following experimental design:

	B1 80-degree room	B2 90-degree room	B3 100-degree room
A1 Low Humidity	Sample 1	Sample 2	Sample 3
A2 High Humidity	Sample 4	Sample 5	Sample 6

Notice that the study involves two levels of humidity and three levels of heat, creating a two-by-three matrix with a total of 6 different treatment conditions. Each treatment condition is represented by a cell in the matrix. For an independent-measures research study, a separate sample would be used for each of the six conditions.

The goal for the two-factor ANOVA is to determine whether the mean differences that are observed for the sample data are sufficiently large to conclude that they are <u>significant</u> differences and not simply the result of sampling error. For the example we are considering, the goal is to determine whether different levels of heat and humidity produce significant differences in performance. To evaluate the sample mean differences, a two-factor ANOVA conducts three separate and independent hypothesis tests. The three tests evaluate:

1. <u>The Main Effect for Factor A</u>: The mean differences between the levels of factor A are obtained by computing the overall mean for each row in the matrix. In this example, the main effect of factor A would compare the overall mean performance with high humidity versus the overall mean performance with low humidity.

2. <u>The Main Effect for Factor B</u>: The mean differences between the levels of factor B are obtained by computing the overall mean for each column in the matrix. In this example, the ANOVA would compare the overall mean performance at $80°$ versus $90°$ versus $100°$.

3. <u>The A x B Interaction</u>: Often two factors will "interact" so that specific combinations of the two factors produce results (mean differences) that are not explained by the overall effects of either factor. For example, changes in humidity (factor A) may have a relatively small overall effect on performance when the temperature is low. However, when the temperature is high ($100°$), the effects of humidity may be exaggerated. In this case, unique combinations of heat and humidity produce results that are not explained by the overall main effects. These "extra" mean differences are the interaction.

The primary advantage of combining two factors (independent variables) in a single research study is that it allows you to examine how the two factors interact with each other. That is, the results will not only show the overall main effects of each factor, but also how unique combinations of the two variables may produce unique results. The interaction can be defined as "extra" mean differences, beyond the main effects of the two factors. An alternative definition is that an interaction exists when the effects of one factor depend on the levels of the second factor.

The Two-Factor Analysis: Each of the three hypothesis tests in a two-factor ANOVA will have its own F-ratio and each F-ratio has the same basic structure

$$F = \frac{\text{variance (differences) between means}}{\text{variance (differences) from error}} = \frac{MS_{between}}{MS_{within}}$$

Each MS value equals SS/df, and the individual SS and df values are computed in a two-stage analysis. The first stage of the analysis is identical to the single-factor ANOVA (Chapter 13) and separates the total variability (SS and df) into two basic components: Between Treatments and Within Treatments. The between-treatments variability measures the magnitude of the mean differences between treatment conditions (the individual cells in the data matrix) and is computed using the basic formulas for $SS_{between}$ and $df_{between}$.

$$SS_{between} = \Sigma\frac{T^2}{n} - \frac{G^2}{N}$$

where the T values (totals) are the cell totals and
n is the number of scores in each cell

$df_{between}$ = the number of cells (totals) minus one

The within-treatments variability measures the magnitude of the differences within each treatment condition (cell) and provides a measure of error variance; that is, unexplained, unpredicted differences due to error.

$$MS_{within} = \frac{SS_{within}}{df_{within}} = \frac{\Sigma SS}{\Sigma df} = \frac{SS_1 + SS_2 + SS_3 + ...}{df_1 + df_2 + df_3 + ...}$$

All three F-ratios use same denominator, MS_{within}

The second stage of the analysis separates the between-treatments variability into the three components that will form the numerators for the three F-ratios: Variance due to factor A, variance due to factor B, and variance due to the interaction. Each of the three variances (MS) measures the differences for a specific set of sample means. The main effect for factor A, for example, will measure the mean differences between rows of the data matrix. The actual formulas for each SS and df are based on the sample totals (rather than the means) and all have the same structure:

$$SS_{between} = \Sigma \frac{T^2}{n} - \frac{G^2}{N}$$

$$= \Sigma \frac{(Total)^2}{number} - \frac{G^2}{N}$$

where the "number" is the number of scores that are summed to obtain each Total

$df_{between}$ = the number of means (or totals) minus one

For factor A, the Totals are the row totals and df equals the number of rows minus 1.

For factor B, the Totals are the column totals and df equals the number of columns minus 1.

The interaction measures the "extra" mean differences that exist after the main effects for factor A and factor B have been considered. The SS and df values for the interaction are found by subtraction.

$$SS_{AxB} = SS_{bet\ cells} - SS_A - SS_B$$

$$df_{AxB} = df_{bet\ cells} - df_A - df_B$$

LEARNING OBJECTIVES

1. Be able to conduct a two-factor ANOVA to evaluate the data from an independent-measures experiment that uses two independent variables.

2. Understand the definition of an interaction between two factors, and be able to recognize an interaction from a description or a graph of experimental results.

The following terms were introduced in this chapter. You should be able to define or describe each term and, where appropriate, describe how each term is related to other terms in the list.

two-factor study	A research study examining two factors (two independent or quasi-independent variables.)
matrix and cells	A two-dimensional table is a matrix and each box in the table is called a cell.
main effect	The overall mean differences between the levels of one factor. When the data are organized in a matrix, the main effects are the mean differences among the rows (or among the columns).
interaction	Mean differences that cannot be explained by the main effects of the two factors. An interaction exists when the effects of one factor depend on the levels of the second factor.

========
========

NEW FORMULAS

========
========

$$SS_{total} = \Sigma X^2 - \frac{G^2}{N} \qquad df_{total} = N - 1$$

$$SS_{\text{bet. cells}} = \Sigma \frac{T^2}{n} - \frac{G^2}{N} \qquad df_{\text{bet. cells}} = (\text{number of cells} - 1)$$

$$SS_{\text{within}} = \Sigma SS_{\text{each cell}} \qquad df_{\text{within}} = \Sigma df_{\text{each cell}}$$

$$SS_{\text{factor A}} = \Sigma \frac{T_A^2}{n_A} - \frac{G^2}{N} \qquad df_{\text{factor A}} = (\text{number levels of A}) - 1$$

$$SS_{\text{factor B}} = \Sigma \frac{T_B^2}{n_B} - \frac{G^2}{N} \qquad df_{\text{factor B}} = (\text{number levels of B}) - 1$$

$$SS_{\text{AxB}} = SS_{\text{bet cells}} - SS_A - SS_B$$

$$df_{\text{AxB}} = df_{\text{bet. cells}} - df_A - df_B$$

STEP BY STEP

Two-Factor ANOVA: The two-factor analysis of variance is used to evaluate mean differences in a research study that uses two independent variables or two quasi-independent variables . The two factors are generally identified as A and B, and the data are presented in a matrix with the levels of factor A determining the rows and the levels of factor B determining the columns. The analysis of variance evaluates three separate hypotheses: one concerning the main effect of factor A, one concerning the main effect of factor B, and one concerning the interaction. In this chapter we considered the two-factor analysis for an independent-measures research study which means that the data consist of a separate sample for each AB treatment combination. The following example will be used to demonstrate the two-factor ANOVA.

The following data are from a two-factor experiment with 2 levels of factor A and 3 levels of factor B. There are $n = 10$ subjects in each treatment condition.

	B1	B2	B3	
A1	T = 10 SS = 20	T = 20 SS = 32	T = 30 SS = 35	$T_{A1} = 60$
A2	T = 10 SS = 15	T = 10 SS = 35	T = 10 SS = 25	$T_{A2} = 30$

$$T_{B1} = 20 \quad T_{B2} = 30 \quad T_{B3} = 40$$

$$G = 90$$
$$\Sigma X^2 = 332$$

Step 1: State the hypotheses and select an alpha level. Because the two-factor ANOVA evaluates three separate hypotheses, there will be three null hypotheses.

For Factor A: H_0: $\mu_{A1} = \mu_{A2}$ (no A-effect)
$\quad\quad\quad\quad\quad$ H_1: $\mu_{A1} \neq \mu_{A2}$

For Factor B: H_0: $\mu_{B1} = \mu_{B2} = \mu_{B3}$ (no B-effect)
$\quad\quad\quad\quad\quad$ H_1: At least one of the B means is different from the others

For AxB: H_0: There is no interaction between factors A and B. That is, the effect of either factor does not depend on the levels of the other factor.
$\quad\quad\quad\quad\quad$ H_1: There is an A x B interaction

Chapter 14 - page 210

We will use $\alpha = .05$ for all three tests.

Step 2: Locate the critical regions. Because there are three separate tests, each with its own F-ratio, we will need to determine the critical region for each test separately. We begin by analyzing the degrees of freedom for these data to determine df for each F. The analysis proceeds in two stages.

$df_{total} = N - 1 = 60 - 1 = 59$
$df_{bet\ cells} = (\text{number of cells}) - 1 = 6 - 1 = 5$
$df_{within} = \Sigma(n - 1) = 9 + 9 + 9 + 9 + 9 + 9 = 54$

This completes the first stage of the analysis. (Check to be certain that the two components add to the total.)

Continuing with the second stage,

$df_A = (\text{number of levels of Factor A}) - 1 = 2 - 1 = 1$
$df_B = (\text{number of levels of Factor B}) - 1 = 3 - 1 = 2$
$df_{AxB} = df_{between\ cells} - df_A - df_B) = 5 - 1 - 2 = 2$

Again, check to be certain that the three components from the second stage add to $df_{bet\ cells}$.

For these data, factor A will have an F-ratio with df = 1, 54. Factor B and the AxB interaction both will have F- ratios with df = 2, 54. Thus, we need F distributions and critical regions for two separate df values. Sketch the two distributions and locate the extreme 5% in each. (Because df = 54 is not listed, we have used 55 for the denominator in each case.)

F distribution for testing the main effect of Factor A.

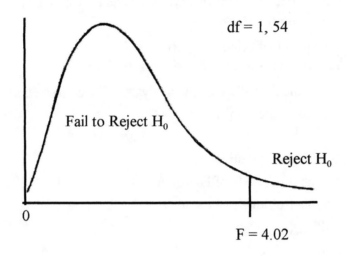

df = 1, 54

Fail to Reject H$_0$

Reject H$_0$

0

F = 4.02

F distribution for testing the main effect for Factor B and the AxB interaction.

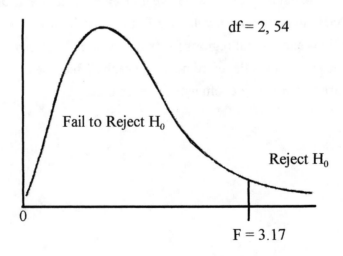

df = 2, 54

Fail to Reject H$_0$

Reject H$_0$

0

F = 3.17

Step 3: Calculate the test statistic. Again, we will need three separate F-ratios. We already have analyzed the degrees of freedom for these data, so we will continue with the analysis of SS. As before, the analysis proceeds in two stages.

$$SS_{total} = \Sigma X^2 - \frac{G^2}{N}$$

$$= 332 - \frac{90^2}{60} = 332 - 135 = 197$$

$$SS_{bet.\ cells} = \Sigma \frac{T^2}{n} - \frac{G^2}{N}$$

$$= \frac{10^2}{10} + \frac{20^2}{10} + \frac{30^2}{10} + \frac{10^2}{10} + \frac{10^2}{10} + \frac{10^2}{10} - \frac{90^2}{60}$$

$$= 10 + 40 + 90 + 10 + 10 + 10 - 135$$

$$= 35$$

$$SS_{within} = \Sigma SS = 20 + 32 + 35 + 15 + 35 + 25$$

$$= 162$$

This completes stage one. Be sure that the two components add to the total.

$$SS_{total} = SS_{bet.\ cells} + SS_{within}$$
$$197 = \quad 35 = 167$$

For the second stage,

$$SS_A = \Sigma \frac{T_A^2}{n_A} - \frac{G^2}{N}$$

$$= \frac{60^2}{30} + \frac{30^2}{30} - \frac{90^2}{60}$$

$$= 120 + 30 - 135$$

$$= 15$$

$$SS_B = \Sigma \frac{T_B^2}{n_B} - \frac{G^2}{N}$$

$$= \frac{20^2}{20} + \frac{30^2}{20} + \frac{40^2}{20} - \frac{90^2}{60}$$

$$= 20 + 45 + 80 - 135$$

$$= 10$$

$$SS_{AxB} = SS_{bet\ cells} - SS_A - SS_B$$

$$= 35 - 15 - 10$$

$$= 10$$

Again, check that these three components from stage two add to $SS_{bet\ cells}$.

Next, compute the MS values that will become the numerators for the three F-ratios.

$$MS_A = \frac{SS_A}{df_A} = 15/1 = 15$$

$$MS_B = \frac{SS_B}{df_B} = 10/2 = 5$$

$$MS_{AxB} = \frac{SS_{AxB}}{df_{AxB}} = 10/2 = 5$$

All three F-ratios will have the same error term denominator:

$$MS_{within} = \frac{SS_{within}}{df_{within}} = 162/54 = 3$$

Finally, the three F-ratios are,

For Factor A: $F = \dfrac{MS_A}{MS_{within}} = 15/3 = 5.00$

For Factor B: $F = \dfrac{MS_B}{MS_{within}} = 5/3 = 1.67$

For AxB: $F = \dfrac{MS_{AxB}}{MS_{within}} = 5/3 = 1.67$

Step 4: Make decision. The F-ratio for Factor A is in the critical region. Therefore, we reject this H_0 and conclude that there is a significant difference between the mean for A1 and the mean for A2. The F-ratios for factor B and for the AxB interaction are not in the critical region. Therefore, we conclude that there is no significant main effect for factor B, and the data are not sufficient to conclude that there is an interaction between factors A and B.

HINTS AND CAUTIONS

1. You should note that several of the SS formulas in the two-factor ANOVA have the same basic structure. Recognizing this structure can make it much easier to learn the formulas. For example, three of the SS formulas are computing variability due to differences between "things." These "things" and the corresponding SS values are:

SS_A (between levels of factor A)
SS_B (between levels of factor B)
$SS_{bet\ cells}$ (between treatment conditions or cells)

The first term of each SS formula involves squaring a total and dividing by the number of scores that were added to compute the total. For example, SS_A squares each of the A totals and divides by bn which is the number of scores used to find each A total. The second term in each of these SS formulas is G^2/N. Thus, all three of these formulas have the same structure that was used to compute SS between treatments for the single-factor ANOVA:

$$SS_{between} = \Sigma \frac{T^2}{n} - \frac{G^2}{N}$$

Note: You also could consider SS_{total} as measuring differences between scores. In this case each score is its own total, and $n = 1$, so the formula for SS_{total} also fits this same general structure. Also note that the degrees of freedom associated with each of these SS values can be determined by simply counting the number of "things" (or totals) and subtracting 1.

2. Remember that the F-ratios for factor A, factor B, and the AxB interaction can all have different values for df and therefore may have different critical values. Be sure that you use the appropriate critical region for each individual F-ratio.

SELF-TEST

True/False Questions

1. In analysis of variance, an independent variable is called a level.

2. A two-factor analysis of variance involves three separate hypothesis tests.

3. In a two-factor ANOVA, a significant interaction means that one of the factors has a significant effect but the second factor does not.

4. In a two-factor ANOVA, all of the F-ratios use the same denominator.

5. If the F-ratios for factor A and factor B both have $df = 1, 36$, then the F-ratio for the interaction will also have $df = 1, 36$.

6. If the F-ratios for factor A and factor B both have $df = 2, 36$, then the F-ratio for the interaction will also have $df = 2, 36$.

7. In a line graph showing the means from a two-factor experiment, if the lines are all straight (not bent), then there is an interaction between the two factors.

8. In a two-factor ANOVA the value for SS_{AxB} is obtained by subtracting SS_A and SS_B from $SS_{bet\ cells}$.

9. If a researcher expects that the difference between two treatment conditions will be greater for males than it is for females, then the researcher is predicting an <u>interaction</u> between the treatments and gender.

10. Whenever a two-factor experiment results in a significant interaction, you should be cautious about interpreting the main effects because an interaction can distort, conceal, or exaggerate the main effects of the individual factors.

Multiple-Choice Questions

1. For an experiment involving 3 levels of factor A and 4 levels of factor B with a sample of n = 5 in each treatment condition, what is the value for df_{within}?
   a. 12
   b. 24
   c. 48
   d. 60

2. The results of a two-factor analysis of variance produce $SS_{bet.\ cells}$ = 12 and SS_{within} = 20. The analysis also produces SS_A = 5 and SS_B = 3. Based on this information, the value of SS_{AxB} would be
   a. 4
   b. 8
   c. 12
   d. Cannot be determined from the information provided

3. For an experiment involving 2 levels of factor A and 3 levels of factor B with a sample of n = 10 in each treatment condition, what are the df values for the F-ratio for factor A?

    a. 1, 9

    b. 2, 9

    c. 1, 54

    d. 2, 54

4. For an experiment involving 2 levels of factor A and 3 levels of factor B with a sample of n = 6 in each treatment condition, what are the df values for the F-ratio for the interaction?

    a. 2, 30

    b. 3, 30

    c. 5, 30

    d. 6, 30

5. The following data represent the means for each treatment condition in a two factor experiment. Note that one mean is not given. What value for the missing mean would result in no main effect for factor A?

    a. 10

    b. 20

    c. 30

    d. 40

	B1	B2
A1	40	30
A2	30	?

6. The following data represent the means for each treatment condition in a two factor experiment. Note that one mean is not given. What value for the missing mean would result in no AxB interaction?

a. 10

b. 20

c. 30

d. 40

	B1	B2
A1	40	30
A2	20	?

7. A two-factor, independent-measures research study is evaluated using an analysis of variance. The F-ratio for factor A has $df = 2, 36$ and the F-ratio for factor B has $df = 3, 36$. Based on this information, what are the df values for the AxB interaction?

a. $df = 5, 36$

b. $df = 6, 36$

c. $df = 5, 72$

d. $df = 6, 72$

8. The results from a two-factor analysis of variance show a significant main effect for factor A and a significant main effect for factor B. Based on this information you can conclude

a. There must by a significant interaction

b. The interaction cannot be significant

c. You cannot make any conclusion about the significance of the interaction

9. The results from a two-factor analysis of variance show that there is no main effect for factor A and there is no main effect for factor B. Based on this information you can conclude

a. There must by a significant interaction

b. The interaction cannot be significant

c. You cannot make any conclusion about the significance of the interaction

10. A two-factor research study is used to evaluate the effectiveness of a new blood-pressure medication. In this two-factor study, Factor A is medication versus no medication and factor B is male versus female. The medicine is expected to reduce blood pressure for both males and females, but it is expected to have a much greater effect for males. This expectation should result in

    a. A significant main effect for factor A (medication)

    b. A significant interaction

    c. All of the above

    d. None of the above

Other Questions

1. Use a two-factor analysis of variance to evaluate the following data from an independent-measures experimental design using $n = 5$ subjects for each treatment condition (each cell). Use $\alpha = .05$ for all tests.

	B1	B2	B3
A1	$\bar{X} = 1$ $T = 5$ $SS = 15$	$\bar{X} = 1$ $T = 5$ $SS = 15$	$\bar{X} = 4$ $T = 20$ $SS = 25$
A2	$\bar{X} = 1$ $T = 5$ $SS = 15$	$\bar{X} = 3$ $T = 15$ $SS = 25$	$\bar{X} = 8$ $T = 40$ $SS = 25$

$$N = 30$$
$$G = 90$$
$$\Sigma X^2 = 580$$

2. The results from a two-factor research study with 2 levels of factor A, 3 levels of factor B, and n = 5 subjects in each treatment condition were evaluated with an analysis of variance. The results are summarized in the following table. Fill in all missing values.

Source	SS	df	MS	
Between Cells	35	__		
Factor A	__	__	__	$F_{(1, 24)} =$ ___
Factor B	20	__	__	$F_{(2, 24)} =$ ___
AxB	__	__	5	$F_{(2, 24)} =$ ___
Within Cells	__	__	2	
Total	__	__		

ANSWERS TO SELF-TEST

True/False Answers

1. False. In ANOVA an independent variable is called a factor.

2. True

3. False. The significance of the interaction is completely independent of the significance of the main effects.

4. True

5. True.

6. False. There are 3 levels for both factors, which makes a total of 9 cells in the matrix. With this structure, $df_{AxB} = 4$, and the F-ratio for the interaction would have $df = 4, 36$.

7. False. An interaction exists if the lines in the graph are not parallel.

8. True

9. True

10. True

Multiple-Choice Answers

1. c   2. a   3. c   4. a   5. d   6. a   7. b   8. c   9. c   10. c

Other Answers

1. The results of the two-factor ANOVA are summarized as follows:

Source	SS	df	MS	
Between Cells	190	5		
Factor A	30	1	30	$F(1, 24) = 6.00$
Factor B	140	2	70	$F(2, 24) = 14.00$
AxB	20	2	10	$F(2, 24) = 2.00$
Within Cells	120	24	5	
Total	310	29		

For $df = 1, 24$ the critical value is 4.26. The F-ratio for factor A is in the critical region so there is a significant difference among the levels of factor A. With $df = 2, 24$ the critical value is 3.40. The F-ratio for factor B is in the critical region so there are significant differences among the levels of factor B. The interaction is not significant.

2.

Source	SS	df	MS	
Between Cells	35	5		
Factor A	5	1	5	$F_{(1, 24)} = 2.50$
Factor B	20	2	10	$F_{(2, 24)} = 5.00$
AxB	10	2	5	$F_{(2, 24)} = 2.50$
Within Cells	48	24	2	
Total	83	29		

Chapter 15

Correlation and Regression

CHAPTER SUMMARY

A correlation is a statistical method used to measure and describe the relationship between two variables. A relationship exists when two variables vary together, as though they are linked. In general, correlational methods measure how much linkage there is. Do the variables always vary together? Sometimes vary together? Perhaps they vary independently of one another, in which case there is no relationship.

A correlation typically evaluates three aspects of the relationship: the direction, the form, and the degree of relationship. The most commonly used correlation is the Pearson correlation (r) which measures the direction and degree of linear relationship. Regression is a statistical procedure that determines the equation for the best-fitting straight line for a set of data. The equation is used to predict Y values from a set of X values.

Usually, correlations are used in situations where a researcher is not attempting to manipulate either of the two variables, but rather is simply observing the variables. To compute a correlation you need two scores, X and Y, for each individual in the sample. The Pearson correlation requires that the scores be numerical values on an interval or ratio scale of measurement. The Spearman correlation requires that the scores are ranks from an ordinal scale of measurement. Other correlational methods exist for other scales of measurement.

The Pearson Correlation

The Pearson correlation requires that you first measure the variability of X and Y scores separately by computing SS for the scores of each variable (SS_X and SS_Y). Then, the covariability (tendency for X and Y to vary together) is measured by the sum of products (SP). The Pearson correlation is found by computing the ratio, $SP/\sqrt{(SS_X)(SS_Y)}$. Thus the Pearson correlation is comparing the amount of covariability (variation from the relationship between X and Y) to the amount X and Y vary separately. The magnitude of the Pearson correlation ranges from 0 (indicating no linear relationship between X and Y) to 1.00 (indicating a perfect straight-line relationship between X and Y). The correlation can be either positive or negative depending on the direction of the relationship.

The Spearman Correlation

The Spearman correlation can be viewed as an alternative to the Pearson correlation. The Pearson correlation measures the degree and direction of <u>linear</u> relationship and the Spearman correlation simply measures the degree and direction of relationship <u>independent of the specific form</u>. Thus, the Spearman correlation can be used when a researcher is not concerned about the exact form of the relationship but simply wants to evaluate the consistency of the relationship. The calculation of the Spearman correlation requires:

1. Two variables are observed for each individual.

2. The observations for each variable are rank ordered. Note that the X values and the Y values are ranked separately.

3. After the variables have been ranked, the Spearman correlation is computed by either:
   a. Using the Pearson formula with the ranked data.
   b. Using the special Spearman formula (assuming there are few, if any, tied ranks).

1. Understand the Pearson correlation and what aspects of a relationship it measures.

2. Know the uses and limitations of measures of correlation.

3. Be able to compute the Pearson correlation by the regular formula (using either the definitional or computational formula for SP) or by the z-score formula.

4. Be able to use a sample correlation to evaluate a hypothesis about the correlation for the general population.

5. Understand the Spearman correlation and how it differs from the Pearson correlation in terms of the data it uses and the type of relationship it measures.

6. Recognize the general form of a linear equation and be able to identify its slope and Y-intercept.

7. Be able to compute the linear regression equation for a set of data.

8. Be able to use the regression equation to compute a predicted value of Y for any given value of X.

NEW TERMS AND CONCEPTS

The following terms were introduced in this chapter. You should be able to define or describe each term and, where appropriate, describe how each term is related to other terms in the list.

positive relationship

A relationship between two variables where increases in one variable tend to be accompanied by increases in the other variable.

negative relationship

A relationship between two variables where increases in one variable tend to be accompanied by decreases in the other variable.

perfect relationship

A relationship where the actual data points perfectly fit the specific form being measured. For a Pearson correlation, the data points fit perfectly on a straight line.

sum of products (of deviations)

A measure of the degree of covariability between two variables; the degree to which they vary together.

Pearson correlation

A measure of the direction and degree of linear relationship between two variables.

significance of a correlation

A demonstration with a hypothesis test showing that a sample correlation is larger than would be expected by chance.

coefficient of determination	The degree to the variability in one variable can be predicted by its relationship with another variable: measured by r^2.
linear relationship	A relationship between two variables where a specific increase in one variable is always accompanied by a specific increase (or decrease) in the other variable.
linear equation	An equation of the form $Y = aX + b$ expressing the relationship between two variables X and Y.
slope	The amount of change in Y for each 1-point increase in X. The value of a in the linear equation.
Y-intercept	The value of Y when $X = 0$. In the linear equation, the value of b.
regression equation for Y	The equation for the best-fitting straight line to describe the relationship between X and Y.
Spearman correlation	A correlation calculated for ordinal data. Also used to measure the consistency of direction for a relationship.
monotonic relation	A relationship that is consistently one-directional.

NEW FORMULAS

$$SP = \Sigma XY - \frac{(\Sigma X)(\Sigma Y)}{n} \quad \text{or} \quad SP = \Sigma(X - \overline{X})(Y - \overline{Y})$$

$$r = \frac{SP}{\sqrt{(SS_X)(SS_Y)}}$$

$$\hat{Y} = bX + a \qquad b = \frac{SP}{SS_X} \qquad a = \overline{Y} - b\overline{X}$$

$$\text{Spearman } r_S = 1 - \frac{6\Sigma D^2}{n(n^2 - 1)}$$

STEP BY STEP

The following example will be used to demonstrate the calculation of the Pearson correlation and the regression equation.

A researcher has pairs of scores (X and Y values) for a sample of $n = 5$ subjects. The data are as follows:

Person	X	Y
#1	0	-2
#2	2	-5
#3	8	14
#4	6	3
#5	4	0

Step 1: Sketch a scatterplot of the data and make a preliminary estimate of the correlation. Also, sketch a line through the middle of the data points and note the slope and Y-intercept of the line.

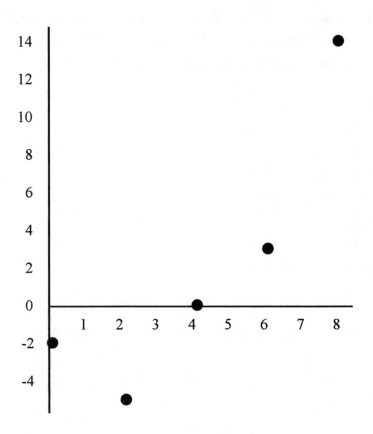

For these data, there appears to be a fairly good, positive correlation - probably around $r = +.7$ or $+.8$. The line has a positive slope and appears to intersect the Y-axis about 5 points below zero.

Step 2: To compute the Pearson correlation you must find SS for both X and Y as well as SP. These same values are needed to find the regression equation. If both sets of scores (X and Y) have means that are whole numbers, then you may use the definitional formulas for SS and SP. Otherwise, it is better to use the computational formulas. Although these data have $\overline{X} = 4$ and $\overline{Y} = 2$, we will demonstrate the computational formulas.

Using one large table, list the X and Y values in the first two columns, then continue with the squared values and the XY products. Find the sum of the numbers in each column. These sums are needed to find SS and SP.

X	Y	X^2	Y^2	XY
0	-2	0	4	0
2	-5	4	25	-10
8	14	64	196	112
6	3	36	9	18
4	0	16	0	0
(totals) 20	10	120	234	120

Use the sums from the table to compute SS for X and Y, and SP.

$$\text{For X: } SS = \Sigma X^2 - \frac{(\Sigma X)^2}{n} = 120 - \frac{20^2}{5}$$

$$= 120 - 80$$

$$= 40$$

For Y: $SS = \Sigma Y^2 - \dfrac{(\Sigma Y)^2}{n}$ $= 234 - \dfrac{10^2}{5}$

$$= 234 - 20$$

$$= 214$$

and $SP = \Sigma XY - \dfrac{(\Sigma X)(\Sigma Y)}{n}$ $= 120 - \dfrac{(20)(10)}{5}$

$$= 120 - 40$$

$$= 80$$

Step 3: Compute the Pearson correlation and compare the answer with your preliminary estimate from Step 1. For these data,

$$r = \frac{SP}{\sqrt{(SS_X)(SS_Y)}} = \frac{80}{\sqrt{(40)(214)}} = \frac{80}{92.52} = 0.865$$

The obtained correlation, $r = +0.865$ agrees with our preliminary estimate.

Step 4: Compute the values for the regression equation, and compare the obtained equation with the preliminary estimates made in Step 1. The general form of the regression equation is, $\hat{Y} = bX + a$. For these data,

$$b = \frac{SP}{SS_X} = \frac{80}{40} = 2$$

and $a = \bar{Y} - b\bar{X}$

$$= 2 - 2(4)$$

$$= -6$$

The regression equation is: $\hat{Y} = 2X - 6$

Both the slope constant (b = +2) and the Y-intercept (a = -6) agree with our preliminary estimates.

The following example demonstrates the calculation of the Spearman Correlation: The Spearman correlation measures the degree of relationship between two variables that are both measured on ordinal scales. If the original data are from interval or ratio scales, you can rank the scores, then compute the Spearman correlation. The following data will be used to demonstrate the calculation of the Spearman correlation.

X	Y
5	12
7	18	(Both X and Y are measured on
2	9	interval scales)
15	14
10	13

Step 1: Check that the X values and the Y values consist of ranks (ordinal data). If not, rank the X's and rank the Y's. Caution: Rank X and Y separately.

For these data,

X Score	X Rank	Y Rank	Y Score
5	2	2	12
7	3	5	18
2	1	1	9
15	5	4	14
10	4	3	13

Step 2: To use the special Spearman formula, compute the difference (D) between the X rank and the Y rank for each individual. Also, find the squared difference (D^2) and the sum of the squared differences.

(Note: The signs of the difference scores are unimportant because you are squaring each D.)

X Rank	Y Rank	D	D^2
2	2	0	0
3	5	2	4
1	1	0	0
5	4	1	1
4	3	1	1
			$6 = \Sigma D^2$

Step 3: Substitute ΣD^2 and n in the Spearman formula.

$$r_S = 1 - \frac{6\Sigma D^2}{n(n^2 - 1)}$$

$$= 1 - \frac{6(6)}{5(25 - 1)}$$

$$= 1 - \frac{36}{120}$$

$$= 0.70$$

There is a positive relation between X and Y for these data. The correlation is fairly high (although not perfect) which indicates a very consistent positive relation.

1. Remember, a correlation of -1.00 indicates a perfect fit, too. The sign indicates the direction of the relationship, not its magnitude.

2. Remember that n refers to the number of individuals (pairs of scores).

3. The formula for the Y-intercept (a) in the regression equation may be easier to remember if you note that this formula simply guarantees that the point defined by $(\overline{X}, \overline{Y})$ is on the regression line. Thus, when $X = \overline{X}$, the predicted Y score will be $\overline{Y}$. In the equation, $\overline{Y} = b\overline{X} + a$. Solving for this equation for the value of a yields, $a = \overline{Y} - b\overline{X}$

4. Remember that a regression equation should not be used to predict values outside the range of the original data.

5. The special formula for the Spearman correlation often causes trouble. Remember, the value of the fraction is computed separately and then subtracted from 1.00. The 1 is not a part of the fraction.

SELF-TEST

True/False Questions

1. A negative correlation indicates that decreases in one variable tend to be accompanied by decreases in the other variable.

2. In a scatterplot, the more the data points vary from a straight line, the closer the value for the Pearson correlation is to zero.

3. For a Pearson correlation, the degree to which the data points fit on a straight line is measured by a numerical value ranging from 0 to 1.00.

4. The value of sum of products(SP) can never be less than zero.

5. A researcher observing children on a school playground noticed that the number of aggressive acts tends to increase as the temperature increases. This is an example of a positive correlation.

6. One or two extreme scores can have a large effect on the value of the Pearson correlation.

7. A Pearson correlation of $r = -1.00$ indicates that the data points fit perfectly on a straight line.

8. In the linear equation $Y = aX + b$, the value of a is called the slope constant.

9. The Spearman correlation is computed from ranks for X and Y.

10. The Spearman correlation will always be a positive value between zero and one

Multiple-Choice Questions

1. In general, the more square-footage a person's home has, the greater the value of the car he/she drives. This demonstrates
    a. a positive relationship
    b. a negative relationship
    c. a cause-effect relationship
    d. all of the above

2. Which of the following Pearson correlations shows the largest magnitude of relationship?

    a. -0.90

    b. +0.74

    c. +0.85

    d. -0.33

3. A set of $n = 5$ pairs of X and Y scores has $\Sigma X = 15$, $\Sigma Y = 5$, and $\Sigma XY = 10$. For these data, the value of SP is

    a. -5

    b. 5

    c. 25

    d. 85

4. The value of the Pearson correlation will be negative if

    a. The value of SS_X is negative

    b. The value of SS_Y is negative

    c. Either SS_X or SS_Y is negative, but not both

    d. SP is negative

5. A set of $n = 15$ pairs of scores (X and Y values) has $SS_X = 4$, $SS_Y = 16$, and $SP = 4$. The Pearson correlation for these data is

    a. $4/20 = 0.20$

    b. $4/64 = 0.0625$

    c. $4/\sqrt{20} = 0.89$

    d. $4/\sqrt{64} = 0.50$

6. A set of n = 15 pairs of scores (X and Y values) produces a correlation of r = 0.85. If 5 points are added to each of the X values and the correlation is computed for the new scores, what value will be obtained for the new correlation?
   a. r = 0.70
   b. r = 0.80
   c. r = 0.85
   d. r = 0.90

7. A correlation is computed for a sample of n = 15 pairs of X and Y values. How large a correlation is necessary to be statistically significant at the .05 level assuming a two-tailed test?
   a. 0.468
   b. 0.482
   c. 0.497
   d. 0.514

8. Which of the following sets of X and Y values (X,Y) is not on the line defined by the equation Y = 3X - 4?
   a. 0, -4
   b. 1, -1
   c. 2, -2
   d. 3, 5

9. The Spearman correlation is used to measure
   a. The degree of linear relationship for ranked data
   b. The degree of one-directional (monotonic) relationship for scores
   c. All of the above
   d. None of the above

10. A set of $n = 5$ pairs of X and Y scores has $SS_X = 4$, $SS_Y = 10$, $SP = 8$, $\overline{X} = 3$, and $\overline{Y} = 10$. Based on this information, the regression equation for predicting Y from X is

    a. $\hat{Y} = 2X + 3$

    b. $\hat{Y} = 8X - 14$

    c. $\hat{Y} = 0.8X + 7.6$

    d. $\hat{Y} = 2X + 4$

Other Questions

1. Compute SP for the following sets of data. You should find that the definitional formula works well with Set I because both means are whole numbers. However, the computational formula is better with Set II where the means are fractions.

Data Set I			Data Set II	
X	Y		X	Y
1	5		1	0
5	2		4	4
6	9		3	1
15	20		2	1
8	4			

2. For the following set of scores

    a. Compute the Pearson correlation.

    b. Find the regression equation for predicting Y from X.

X	Y
0	16
1	6
2	9
3	0
4	9

3.   Find the regression equation for the following set of data.

X	Y
4	1
7	16
3	4
5	7
6	7

4. Suppose that a sample of $n = 42$ pairs of X and Y scores yields a Pearson correlation of $r = +0.40$. Does this sample provide sufficient evidence to conclude that a significant correlation exists in the population? Test at the .05 level of significance, two tails.

5. Compute the Spearman correlation for each of the following sets of data. (Note that you will need to rank order the X and Y values for the data in Set 2.)

Set 1			Set 2	
X and Y measured on ordinal scales			X and Y measured on interval scales	
X	Y		X	Y
2	5		1	5
4	1		5	2
3	2		6	9
1	4		15	20
5	3		8	4

ANSWERS TO SELF-TEST

True/False Answers

1. False. For a negative correlation, decreases in one variable tend to be accompanied by increases in the other variable.

2. True

3. True

4. False. The value of SP can be either positive or negative.

5. True

6. True

7. True

8. True

9. True

10. False. The value of the Spearman correlation can range in value from +1.00 to -1.00.

Multiple-Choice Answers

1. a   2. a   3. a   4. d   5. d   6. c   7. d   8. c   9. c   10. d

Other Answers

1. For data set I, SP = 121. For data set II, SP = 6.

2. a. $SS_X = 10$, $SS_Y = 134$, and SP = -20. The Pearson correlation is r = -0.546.

    b. The regression equation is, $\hat{Y} = -2X + 12$.

3. For these data, $SS_X = 10$ and SP = 30. The regression equation is,
    $\hat{Y} = 3X - 8$

4. The null hypothesis states that there is no relationship in the population. H_0: $\rho = 0$ With $n = 42$, the correlation must be greater than 0.304 to be significant. This sample correlation is sufficient to conclude that there is a significant correlation in the population.

5. For data set 1, the Spearman correlation is r_s = -0.60. After ranking the scores in data set 2, the Spearman correlation is r_s = +0.50.

Chapter 16

Hypothesis Tests with
Chi-Square

===============

CHAPTER SUMMARY
===============

Chapter 16 introduces two non-parametric hypothesis tests using the chi-square statistic: the chi-square test for goodness of fit and the chi-square test for independence. The term "non-parametric" refers to the fact that the chi-square tests do not require assumptions about population parameters nor do they test hypotheses about population parameters. The most obvious difference between the chi-square tests and the other hypothesis tests we have considered (t and ANOVA) is the nature of the data. For chi-square, the data are frequencies rather than numerical scores.

The test for goodness-of-fit uses frequency data from a sample to test hypotheses about the shape or proportions of a population. Each individual in the sample is classified into one category on the scale of measurement. The data, called observed frequencies, simply count how many individuals from the sample are in each category. The null hypothesis specifies the proportion of the population that should be in each category.

The test for independence can be used and interpreted in two different ways:

1. Testing hypotheses about the relationship between two variables in a population, or

2. Testing hypotheses about differences between population proportions.

Although the two versions of the test for independence appear to be different, they are equivalent and they are interchangeable. The first version of the test emphasizes the relation between chi-square and a correlation because both procedures examine the relationship between two variables. The second version of the test emphasizes the relation between chi-square and an independent-measures t test (or ANOVA) because both tests use data from two (or more) samples to test hypotheses about two (or more) populations.

The first version of the chi-square test for independence requires one sample where each individual is classified on two different variables. The data are usually presented in a matrix with the categories for one variable defining the rows and the categories of the second variable defining the columns. The data, called observed frequencies, simply show how many individuals from the sample are in each cell of the matrix. The null hypothesis for this test states that there is no relationship between the two variables; that is, the two variables are independent.

The second version of the test for independence requires a separate sample for each population being compared. The same variable is measured for each sample by classifying individual subjects into categories of the variable. The data are presented in a matrix with the different samples defining the rows and the categories of the variable defining the columns. The data, again called observed frequencies, show how many individuals are in each cell of the matrix. The null hypothesis for this test states that the proportions (the distribution across categories) are the same for all of the populations.

The calculation of the chi-square statistic require two steps:

1) The null hypothesis is used to construct an idealized sample distribution of <u>expected frequencies</u> that describes how the sample would look if the data were in perfect agreement with the null hypothesis.

For the goodness of fit test, the expected frequency for each category is obtained by

$$\text{expected frequency} = f_e = pn$$

(p is the proportion from the null hypothesis and n is the size of the sample)

For the test for independence, the expected frequency for each cell in the matrix is obtained by

$$\text{expected frequency} = f_e = \frac{(\text{row total})(\text{column total})}{n}$$

2) A chi-square statistic is computed to measure the amount of discrepancy between the ideal sample (expected frequencies from H_0) and the actual sample data (the observed frequencies $= f_o$). A large discrepancy results in a large value for chi-square and indicates that the data do not fit the null hypothesis and the hypothesis should be rejected. The calculation of chi-square is the same for all chi-square tests:

$$\text{chi-square} = \chi^2 = \Sigma \frac{(f_o - f_e)^2}{f_e}$$

The fact that chi-square tests do not require scores from an interval or ratio scale makes these tests a valuable alternative to the t tests, ANOVA, or correlation because they can be used with data measured on a nominal or an ordinal scale.

LEARNING OBJECTIVES

1. Recognize the experimental situations where a chi-square tests is appropriate.

2. Be able to conduct a chi-square test for goodness of fit to evaluate a hypothesis about the shape of a population frequency distribution.

3. Be able to conduct a chi-square test for independence to evaluate a hypothesis about the relationship between two variables.

NEW TERMS AND CONCEPTS

The following terms were introduced in this chapter. You should be able to define or describe each term and, where appropriate, describe how each term is related to other terms in the list.

parametric statistical tests

A test evaluating hypotheses about population parameters and making assumptions about parameters. Also, a test requiring numerical scores.

non-parametric statistical tests

A test that does not test hypotheses about parameters or make assumptions about parameters. The data usually consist of frequencies.

expected frequencies

Hypothetical, ideal frequencies that are predicted from the null hypothesis.

observed frequencies

The actual frequencies that are found in the sample data.

chi-square statistic

A test statistic that evaluates the discrepancy between a set of observed frequencies and a set of expected frequencies.

chi-square distribution	The theoretical distribution of chi-square values that would be obtained if the null hypothesis was true.
chi-square test for goodness of fit	A test that uses the proportions found in sample data to test a hypothesis about the corresponding proportions in the general population.
chi-square test for independence	A test that uses the frequencies found in sample data to test a hypothesis about the relationship between two variables in the population.

NEW FORMULAS

$$\chi^2 = \Sigma \frac{(f_o - f_e)^2}{f_e}$$

$f_e = pn \qquad$ (test for goodness of fit)

$$f_e = \frac{(\text{Row Total})(\text{Column Total})}{n} \qquad \text{(test for independence)}$$

STEP BY STEP

The chi-square test for independence: The chi-square test for independence uses frequency data to test a hypothesis about the relationship between two variables. The null hypothesis states that the two variables are independent (no relationship). Rejecting H_0 indicates that the data provide convincing evidence of a consistent relationship between the two variables. The following example will be used to demonstrate this chi-square test.

A psychologist would like to examine preferences for the different seasons of the year and how these preferences are related to gender. A sample of 200 people is obtained and the individuals are classified by sex and preference. The psychologist would like to know if there is a consistent relationship between sex and preference. The frequency data are as follows:

Favorite Season

	Summer	Fall	Winter	Spring
Males	28	32	15	45
Females	32	8	5	35

Step 1: State the hypotheses and select an alpha level. The null hypothesis says that there is no relationship.

H_0: Preference is independent of sex. One version of the null hypothesis states that there is no relationship between preference and sex. The second version of H_0 states that there is no difference between the distribution of preferences for males and the distribution of preferences for females (both distributions have the same proportions).

Chapter 16 - page 249

The alternative hypothesis simply says that the two variables are not independent.

    H_1: Preference is related to sex, or the distribution of preferences is different for males and females.

We will use $\alpha = .05$

Step 2: Locate the critical region. The degrees of freedom for the chi-square test for independence are

    $df = (C - 1)(R - 1)$

For this example, $df = 3(1) = 3$. Sketch the distribution and locate the extreme 5%. The critical boundary is 7.81.

The χ^2 distribution with $df = 3$

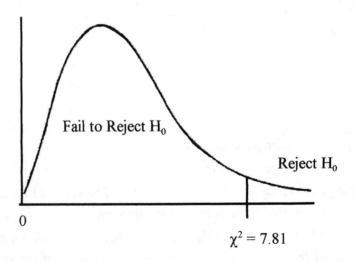

Fail to Reject H_0

Reject H_0

0

$\chi^2 = 7.81$

Step 3: Compute the test statistic. The major concern for this chi-square test is determining the expected frequencies. We begin with a blank matrix showing only the row and column totals from the data.

Favorite Season

	Summer	Fall	Winter	Spring	
Males					80
Females					120
	60	40	20	80	

The expected frequencies are determined by the null hypothesis. In this example, H_0 says that the distribution of preferences is the same for both genders. Therefore, we must determine the "distribution of preferences."

For the total sample of $n = 200$ the data show:

60/200 = 30% prefer Summer
40/200 = 20% prefer Fall
20/200 = 10% prefer Winter
80/200 = 40% prefer Spring

Next, we apply this distribution to each sex group. There are 120 males. Using the proportions from the overall distribution, we would expect,

30% of 120 = 36 males prefer summer
20% of 120 = 24 males prefer fall
10% of 120 = 12 males prefer winter
40% of 120 = 48 males prefer spring

For the group of 80 females we would expect

    30% of 80 = 24 females prefer summer

    20% of 80 = 16 females prefer fall

    10% of 80 = 8 females prefer winter

    40% of 80 = 32 females prefer spring

Place these values in a matrix of expected frequencies.

Favorite Season

	Summer	Fall	Winter	Spring	
Males	36	24	12	48	80
Females	24	16	8	32	120
	60	40	20	80	

Now you are ready to compute the chi-square statistic.

a) For each cell in the matrix, find the difference between the expected and the observed frequency.

b) Square the difference.

c) Divide the squared difference by the expected frequency.

d) Sum the resulting values for each category

f_o	f_e	$(f_o - f_e)$	$(f_o - f_e)^2$	$(f_o - f_e)^2/f_e$
28	36	-8	64	1.78
32	24	8	64	2.67
15	12	3	9	0.75
45	48	-3	9	0.19
32	24	8	64	2.67
8	16	-8	64	4.00
5	8	-3	9	1.13
35	32	3	9	0.28

$$13.47 = \chi^2$$

Step 4: Make decision. The chi-square value is in the critical region. Therefore, we reject H_0 and conclude hat there is a significant relation between sex and preferred season.

To describe the nature of the relationship, you can compare the data with the expected frequencies. From this comparison it should be clear that more men prefer fall and more women prefer summer than would be expected by chance.

HINTS AND CAUTIONS

1. When computing expected frequencies for either chi-square test, it is wise to check your arithmetic by being certain that $\Sigma f_e = \Sigma f_o = n$. In the test for independence, the expected frequencies in any row or column should sum to the same total as the corresponding row or column in the observed frequencies.

2. Whenever a chi-square test has df = 1, the difference (absolute value) between f_o and f_e will be the same for every category. This is true for either the goodness of fit test with 2 categories, or the test of independence with 4 categories. Knowing this fact can help you check the calculation of expected frequencies and it can simplify the calculation of the chi-square statistic.

True/False Questions

1. In a chi-square test, the sample data are called <u>observed frequencies</u>.

2. One advantage of the chi-square tests is that they can be used when the data are measured on a nominal scale.

3. Suppose that the null hypothesis for a chi-square test for goodness of fit predicts no preferences among three categories. If a sample of $n = 60$ subjects is used, then the expected frequency for each category would be $f_e = 20$.

4. A chi-square test for goodness of fit is used to evaluate a hypothesis about how a population is distributed across three categories. If the researcher uses a sample of $n = 100$ subjects, then the chi-square test will have $df = 99$.

5. The chi-square test for independence requires that each individual be categorized on two separate variables.

6. The df value for a chi-square test does <u>not</u> depend on the sample size.

7. In a chi-square test, it is possible for the observed frequencies to be fractions or decimal values.

8. In general, a large value for chi-square will tend to reject the null hypothesis.

9. A chi-square statistic can never have a value less than zero.

10. A researcher is using a chi-square test for independence to evaluate the relationship between birth-order position and self esteem. Each individual is classified as being 1st born, 2nd born, or 3rd born and self-esteem is categorized as either high or low. For this study, the chi-square statistic will have df = 2.

Multiple-Choice Questions

1. For the chi-square test for goodness of fit, what is the value of df for a test with four categories and a sample of n = 100?
    a. 3
    b. 4
    c. 96
    d. 99

2. The null hypothesis for the chi-square test for goodness of fit specifies
    a. Proportions for the entire population
    b. Proportions for the sample
    c. Frequencies for the entire population
    d. Frequencies for the sample

3. The chi-square distribution is
    a. Symmetrical, centered at a value of zero
    b. Symmetrical, centered at a value determined by the degrees of freedom
    c. Positively skewed, with no values less than zero
    d. Negatively skewed, with no values greater than zero

4. A chi-square test for independence is used to evaluate the relationship between two variables. If one variable is classified into 4 categories and the other variable is classified into 2 categories, then the chi-square statistic will have
     a. df = 3
     b. df = 6
     c. df = 8
     d. Cannot be determined from the information provided

5. The sample data for a chi-square test are called
     a. Expected frequencies
     b. Observed frequencies
     c. Expected proportions
     d. Observed proportions

6. A researcher would like to test the claim that 9 out of 10 doctors prefer Brand X. A sample of 60 doctors is obtained and each is asked to compare Brand X with another leading brand. The data show that 48 of the doctors picked Brand X. If these data are evaluated using a chi-square test for goodness of fit, what is the expected frequency for Brand X?
     a. 9
     b. 12
     c. 48
     d. 54

7. A researcher is examining the relationship between color preferences and gender. A sample of 30 men and 30 women is obtained and each person is asked to identify his/her preference between two choices of paint colors for a new student lounge. For this sample, 5 of the men preferred color A, and 15 of the women preferred color A. If a chi-square test is used to evaluate the relationship, what is the expected frequency for men preferring color A?

    a. 5
    b. 10
    c. 15
    d. 20

8. A researcher is examining the relationship between color preferences and gender. A sample of 30 men and 30 women is obtained and each person is asked to identify his/her preference between two choices of paint colors for a new student lounge. For this sample, 5 of the men preferred color A, and 15 of the women preferred color A. If a chi-square test is used to evaluate the relationship, what is the df value for the chi-square statistic?

    a. 1
    b. 3
    c. 57
    d. 59

9. In the chi-square test for independence, the null hypothesis states
    a. There is no difference between the two variables being examined
    b. There is a difference between the two variables being examined
    c. There is no relationship between the two variables being examined
    d. There is a relationship between the two variables being examined

10. A researcher is examining preferences among 5 brands of pizza using a sample of n = 60 individuals. Each individual tastes all 5 brands and selects his/her favorite. If the data are evaluated with a chi-square test for goodness of fit using α = .05, then a significant result would require a chi-square statistic

    a. Greater than 9.49

    b. Less than 9.49

    c. Greater than 79.08

    d. Less than 79.08

Other Questions

1. A researcher is examining preferences among four new flavors of ice cream. A sample of n = 80 people is obtained. Each person tastes all four flavors and then picks his/her favorite. The distribution of preferences is as follows.

Ice Cream Flavor

A	B	C	D
12	18	28	22

Do these data indicate any significant preferences among the four flavors? Test at the .05 level of significance.

2. A researcher is testing four new flavors of bubble gum using a sample that consists of 50 men, 200 women, and 250 children. Each individual selects his/her favorite flavor. In the total sample of 500 people, 100 selected Flavor A, 200 chose B, 150 picked C, and only 50 preferred D.

    a. If these data are used for a chi-square test for independence, state the null hypothesis.

    b. Use the following matrix to fill in the expected frequencies for the chi-square test for independence.

Flavor

	A	B	C	D	
Men					50
Women					200
Children					250
	100	200	150	50	

3. A researcher is interested in the relationship between birth order and personality. A sample of $n = 100$ people is obtained, all of whom grew up in families as one of three children. Each person is given a personality test and the researcher also records the person's birth order position (1st born, 2nd, or 3rd). The frequencies from this study are shown in the following table. On the basis of these data can the researcher conclude that there is a significant relation between birth order and personality? Test at the .05 level of significance.

Birth Position

	1st	2nd	3rd
Outgoing	13	31	16
Reserved	17	19	4

ANSWERS TO SELF-TEST

True/False Answers

    1. True

    2. True

    3. True

4. False. For the chi-square tests, degrees of freedom are determined by the number of categories, not the number of subjects.

5. True.

6. True.

7. False. Observed frequencies are actual counts of how many individuals are in each category. These are always whole numbers.

8. True

9. True

10. True

Multiple-Choice Answers

1. a   2. a   3. c   4. a   5. b   6. d   7. b   8. a   9. c   10. a

Other Answers

1. The null hypothesis states that there are no preferences among the four flavors and than 1/4th of the population should prefer each flavor. The expected frequency are 20 for each flavor. With $df = 3$, the critical value is 7.81. For these data, the chi-square statistic is 6.80. The decision is to fail to reject the null hypothesis and conclude that the data do not provide sufficient evidence to indicate significant preferences among the four ice cream flavors.

2. a. One version of the null hypothesis states that gum preference is independent of the age/gender classifications. The second version of H_0 states that the distribution of flavor preferences is the same (same proportions) for men, women, and children.

b.

	Flavor				
	A	B	C	D	
Men	10	20	15	5	50
Women	40	80	60	20	200
Children	50	100	75	25	250
	100	200	150	50	

3. The null hypothesis states that there is no relation between birth order and personality - the two variables are independent. With $df = 2$, the critical value for this test is 5.99. The expected frequencies are as follows:

	Birth Position		
	1st	2nd	3rd
Outgoing	18	30	12
Reserved	12	20	8

For these data, the chi-square statistic is 6.89. Reject H_0 and conclude that there is a significant relation between personality and birth order.